CHAIN REACTIONS

CHAIN REACTIONS

The Hopeful History of Uranium

LUCY JANE SANTOS

PEGASUS BOOKS
NEW YORK LONDON

CHAIN REACTIONS

Pegasus Books, Ltd.
148 West 37th Street, 13th Floor
New York, NY 10018

Copyright © 2024 by Lucy Jane Santos

First Pegasus Books cloth edition November 2024

ISBN: 978-1-63936-744-3

10 9 8 7 6 5 4 3 2 1

Printed in the United States of America
Distributed by Simon & Schuster
www.pegasusbooks.com

'Nothing in life is to be feared, it is only to be understood. Now is the time to understand more, so that we may fear less.'
Marie Skłodowska Curie

'You know what uranium is, right? It's this thing called nuclear weapons and other things, like lots of other things are done with uranium, including some bad things.'
Donald J. Trump

CONTENTS

PROLOGUE

This story starts just a smidge under 5 billion years ago.[1]

We begin with a solar nebula, a cloud of dust and gas.

Contained within this cloud are various elements. Scientists believe that the heavy elements, which include not only uranium, but also gold and platinum, originate from two neutron stars colliding about 80 million years before the birth of the solar system.

These collisions flung the uranium, along with other stardust, across the universe. For eons they voyaged, carried by interstellar winds and cosmic currents. Gravity doing what it does best caused this matter to condense and eventually form the sun at the centre, as well as the planets, including our own.

Eventually, as smaller rocky worlds like Earth formed through collisions of elemental materials, uranium became an integral part of the planet's composition and one of the most common elements in the Earth's crust. It nestled within rocks, soil and even the oceans. Over vast periods of time, geological forces shaped the Earth's surface, creating mountains and valleys.

Deep within the Earth's crust, buried as layer upon layer of sediment was built up, uranium found its haven in various geological formations, ranging from ancient sedimentary layers to crystalline structures. At the same time, uranium's natural life cycle led to a slow radioactive decay, which is the main heat source within the core of our planet.

In the Oklo Valley in Gabon, Africa, uranium deposits built up in such a specific way that they underwent nuclear fission, a reaction when the nucleus of an atom splits and releases a large amount of energy.

For hundreds of thousands of years these sixteen natural reactors, known collectively as the Oklo Fossil Reactors, hummed and buzzed and generated a modest amount of heat and energy in a self-sustaining reaction before eventually dying out.

Time and geological processes sculpted the Earth further, revealing its hidden treasures. Erosion and weathering gradually unveiled uranium-bearing ores, exposing their vibrant hues and eventually ensuring a worldwide search for the element as scientists and governments sought to first understand and then control its powers.

THE EARLY HISTORY

1

Johannes Kentmann was fascinated by rocks. The Dresden-born physician had other passions – he also wrote books about herbal medicine, as well as biology and botany[1] – but it was mineralogy that stirred him the most. Kentmann was so passionate about his subject that he had a cabinet with thirteen numbered drawers installed to house his comprehensive collection of minerals, which he called an 'ark'. In 1565, this collection was catalogued in *De Omni Rerum Fossilium Genere, Gemmis, Lapidibus, Metallis, et huiusmodi* (*On Every Kind of Fossils, Gems, Stones, Metals and the Like*), a composite volume of works by seven authors edited by the Swiss naturalist Conrad Gessner.

Kentmann's volume – described in Gessner's book as being collated by 'the first man in Europe to make a collection of minerals' – consisted of individual entries for 1,608 specimens representing 26 different groups of minerals from 135 locations. And one of these, the ore bechblende, is the beating black heart in the story of uranium.

Pechblende (somewhere in the fourteenth century there was a shift in German vowels and the b became a p) seems to have been named for both its appearance and for its reputation. As Kentmann's catalogue entry indicates, the substance was black, hence 'pech', which in German can mean dark and sticky. 'Blende' means mixture, which is a fitting description, as pechblende can be made up from up to 30 different elements. But there was a double meaning to the name as well: 'blenden' means to deceive and 'pech haben' is to have bad luck. While Kentmann was the first to publish a formal identification – listing it as a sterile lead similar to black pitch – the mineral wasn't totally unknown.

In fact, miners were very familiar with pechblende, much to their annoyance. Finding pechblende meant the valuable stuff you were looking for, like silver or gold, was almost running out. Pechblende was a sign to move on.

Later, the sheer quantity of pechblende would be a huge problem for miners working in the Erzgebirge mountain range, which ran between the borders of Saxony and Bohemia. But it wasn't always that way; in the early 1500s a rich vein of silver ore had been discovered in the Bohemian part of the Erzgebirge, an area that is now part of the Czech Republic.

The owner of the land, Count Stefan Schlick, built a castle called Schloss Freudenstein to fortify a newly built town to house the miners. By 1518, there were over 400 houses and 8,000 miners working the site, and the town came to be known as Sankt Joachimsthal.[2]

Thanks to the rich pickings in the mine, St Joachimsthal grew in importance, being granted royal status in 1520.[3] As befitting of their new significance, the Schlicks were also given permission to mint their own coins, an honour which allowed them to consolidate the power of the town. Schlick brought in two mint masters, Ulrich Gebhart and Stephan Gemisch, and developed a new currency based on coins known as 'Joachimsthalers'. By the seventeenth century the use of these large silver coins had spread across central Europe and they became the dominant unit of currency, accepted in neighbouring kingdoms without the need to exchange for local monies.[4]

The Schlicks had been appointed as administrators and the principal suppliers of the silver used to mint all the coins. However, these privileges only lasted a few years. The state treasury soon realised they were missing out on the profits from producing their own coins and took the Schlick mint and placed it under royal administration and operation. Nevertheless, the Schlicks still profited greatly from the arrangement, with an estimated 250,000 kilograms of silver mined between 1516 and 1554.[5]

Joachimsthal was a prosperous and vibrant town that attracted many influential people, including Georg Bauer, who was better known by his pen name Georgius Agricola. Bauer was a linguist, scholar and teacher who was appointed as the town's physician in 1527, having obtained his medical degree from the University of Bologna.[6] During his few years in the town, he became very interested in the mines and its workers and went on to publish ten texts on the topic, the first one in 1530.

His best-known work, *De Re Metallica* (*On the Nature of Metals*) was published posthumously. Written in Latin, the book is divided into twelve chapters, each of which deals with a different aspect of mining and metallurgy. It is also known for its vivid, detailed illustrations, made by artists and woodcutters.[7] Some of the images in *De Re Metallica* show miners digging for ore, using hand tools and explosives to extract minerals from the earth. Others show the refining process, such as the smelting of metals in furnaces and the casting of metals into moulds. The book also includes images of various technologies in use at the time, such as bellows to provide air to furnaces and waterwheels to power mills.

Bauer was specifically concerned about the working conditions of the miners and, ultimately, the perils of following profits at all costs warning:

> It remains for me to speak of the ailments and accidents of miners, and of the methods by which they can guard against these, for we should always devote more care to maintaining our health, that we may freely perform our body functions, than to making profits.[8]

While mining was, and still is, an incredibly dangerous occupation in its own right, Bauer was particularly concerned about a respiratory illness known as bergsucht, which had first been described by the alchemist known as Paracelsus in 1533 and referred to as 'Mala Metallorum'.[9]

Although no one knew exactly what caused this often-fatal illness, it was widely accepted that it was caused by some kind of poisonous dust. And while arsenic, one of the many toxic minerals found in the mountains, was the official suspect, the miners had their own theory: evil mountain gnomes who were intent on punishing those who violated their underground domain.[10]

Bauer didn't discount this:

> In some of our mines, however, though in very few, there are other pernicious pests. These are demons of ferocious aspect, about which I have spoken in my book *De animatibus subterrabeis* [On Underground Spirits]. Demons of this kind are expelled and put to flight by prayer and fasting.[11]

Bauer didn't just report on the problems; he also suggested practical ways to improve the conditions within the mines, including masks, gloves and protective clothing.[12] Over fifteen pages Bauer discusses the construction of different types of ventilation machines, using wind, fans and bellows, which could help provide fresh air rather than the stagnant air naturally present in the mines.[13]

Bauer died in 1555 at the age of 61 – his recommendations either not carried out or found ineffective against the dreaded bergsucht. His contemporary Paracelsus was also unable to pinpoint what was ailing the miners, but he did distinguish between the acute and chronic toxic effects of metals.[14]

By the latter half of the sixteenth century, Joachimsthal's prospects were in a downward spiral that it never really recovered from. There were outbreaks of plague, a change of ruling empires, suppression of nationalism, the town was sacked and Freudenstein castle was destroyed by fire. Competition from outside of Europe was also driving the price of silver down, most notably the discovery and exploitation of vast silver mines in Mexico and Peru following the conquests of Cortés and Pizarro. These mines were worked on a massive scale, using forced labour, including African slaves. The Spanish colonisers in Peru even encouraged coca-chewing by the workers to increase their energy and therefore their productivity.[15] And compounding these issues was the fact that the silver in the Joachimsthal mines, or at least the silver that the technology available at the time allowed them to reach, was running out.[16]

Decades after Joachimsthal had begun its freefall into historical obscurity, a man started a journey that would change the town's fortunes once again. Martin Heinrich Klaproth might have set out on the path to priesthood, but fate had other plans for him. Instead of pursuing a religious vocation, Klaproth became a self-taught expert in analysing minerals, investigating hundreds of different samples in his lifetime. And he made some game-changing discoveries along the way. Arguably the founder of the new science of analytic chemistry, he discovered zirconium (1789), separated strontium-28 from calcium and confirmed the discovery of the substance that came to be known as titanium (1792).

But Klaproth's most groundbreaking work came in 1789 while investigating two of the minerals found in Erzgebirge's silver deposits: pechblende and torbernite. What he encountered during his analysis of torbernite led him to a stunning discovery, but it was pechblende that he used to conduct his further investigations.[17] Dissolving the ore in nitric acid and neutralising the solution with sodium hydroxide, he found a yellow compound that he heated with charcoal to obtain a black powder. As his experiments had determined that this sample was chemically indivisible, i.e. he couldn't separate anything else from it, Klaproth was convinced he had discovered a new element.

On the evening of 24 September 1789, Klaproth addressed the Royal Prussian Academy of Sciences in Berlin and reported his results:

> The number of known metals has been increased by one – from 17 to 18. This I have called a metalloid, a new element which I see as a strange kind of half metal. It is not related to the zinc, iron, or tungsten found in so-called pitchblende. For some reason, however, I have found it associated with lead. Consequently, I suggest that past errors in naming should be eradicated – such as iron pitch ore, pitchblende or black tin ore. I have chosen a name. A few years ago we thrilled to hear of the discovery of the final planet by Sir William Herschel. He called this new member of our solar system Uranus. I proposed to borrow from the honour of that great discovery and call this new element – Uranium.[18]

Klaproth's discovery was later called into question in 1841 when Eugène-Melchior Péligot, a professor of analytical chemistry at the École Centrale des Arts et Manufactures in Paris, utilised advancements in chemical analysis and tested the pechblende ore again. He determined that Klaproth had not found a pure element but, instead, uranium and oxygen as a compound – uranium oxide. Péligot was the first to create pure uranium and study its properties, including its atomic mass.

Of course, once something new has been discovered there is often a race to monetise the substance. And uranium had some unique properties that meant it was easily and quickly utilised in all manner of things. However, its greatest use was really in the ceramic and glass industries. And while this use snowballed in the late-nineteenth century, it is also apparent that uranium may have been used on a non-commercial basis as a colouring agent prior to its formal identification.

In particular, the discovery of uranium oxide in a Roman glass mosaic has caused some discussion among glass experts, as it is the earliest known use of uranium oxide in existence.

The mosaic was found in a 1912 archaeological dig carried out by Robert T. Günther, historian at the luxurious villa of Publius Vedius Pollio at Posillipo, an area to the south-west of Naples, overlooking the bay. After the death of Vedius Pollio in 15 BCE,[19] it was given to the Roman emperor Augustus, and was levelled, rebuilt and extended to become an imperial villa. It remained an imperial possession until at least 183 CE.

Fast-forward to the early-twentieth century and the ruin was being worked over by Günther and his team. In one room that was decorated with light-coloured walls, divided into panels by lines painted in black and the colour the Romans called Sinopsis, which is today better known as Pompeiian red, was a small niche encrusted with a mosaic. This portrayed a white dove hovering with wings outstretched and tail spread in the blue sky over green plants. It is a striking image with great attention to detail and was dated by the team to around 79 CE.

After photographing and sketching its position in the room, Günther's team took a few pieces for analysis but left the mosaic in situ. The samples were taken to J.J. Manley, a chemist at Magdalen College, Oxford. The blue tesserae were simply cobalt, a relatively common find, but the green tiles were of more interest. Manley determined that some of them contained trace substances of uranium oxide.[20] Unfortunately the mosaic itself disappeared around the Second World War and the samples were lost and never recovered.[21]

As this is the only example of uranium found in glass of this period it does rather beg the question of how it got there. One possibility, of course, is that it was accidental. In this scenario the Roman glassmaker used sand that just happened to contain trace amounts of uranium. After all, to make glass you just need sand, nitrate and a tremendous amount of heat to fuse everything together. On the other hand, Roman glassmakers knew how to colour glass by adding in metallic oxides, and there are beautiful examples from all periods, including those blue cobalt oxide tiles at Posillipo, so it could have been a deliberate, albeit seemingly one-off experiment.

There is also a hypothesis that the mosaic glass could have originated from Roman Britain, specifically from Cornwall, an area that was a significant mining centre for various metals, including uranium.

We know this from Charles Sandoe Gilbert, a druggist and historian who published *An Historical Survey of the County of Cornwall* in 1817, a comprehensive account of the county's history, geology and natural resources. The book covers a wide range of topics, from pre-history to the nineteenth century, and provides detailed descriptions of Cornwall's towns, villages and notable landmarks.

Cornwall was one of the most important mining centres in Europe, comparable to those in the Erzgebirge mountains, from pre-Roman times until the twentieth century. The region commercially mined several metals, including antimony, arsenic, cobalt and manganese.[22]

The first reference to uranium minerals in Cornwall dates back to 1805, when what is thought to have been torbernite was identified in mine dumps at Tincroft Mine near Camborne.[23] And Gilbert mentions mines at Carharrack, Huel Garland, Tolcarne and Huel Unity.

Gilbert notes that uranium could be used in glassmaking: 'it combines with oxygen, and its oxides impart bright colours to glass, which are according to the proportions brown, apple green, or emerald green'.[24]

While there is no suggestion that uranium glass was being produced commercially or in any significant amount in Cornwall, or indeed anywhere else, at this time, this changed shortly afterwards, with a number of companies exploring the use of uranium glass in the second quarter of the nineteenth century.[25]

So, we see the Harrach glassworks in Saxony exhibiting their yellow-green glass at the Prague Exhibition in 1831.[26] We know, thanks to records held by the Museum of London, that the British glass manufacturers Whitefriars used uranium as a colouring agent from 1836. And it appears to have been a technique that was embraced quickly by manufacturers keen to show off their skill and create new forms of beautiful glassware. Hence the silver mounted candlesticks with prisms of topaz glass coloured with uranium that were presented to Queen Adelaide in 1836.[27] And the twelve finger bowls, also made by Whitefriars, of uranium topaz glass that were used at the Corporation of London Banquet for the new queen, Victoria, held at the Guildhall on 9 November 1837.[28]

Those examples were celebratory, one-off pieces made to welcome queens, not for everyday use. And while, as with many historical advancements, there may be some debate and uncertainty about the exact origins and timelines, the originator of commercially available uranium glass is typically recognised to be the Riedel company, which was founded by Josef Riedel in Bohemia. The Riedel company developed two types of coloured glass using uranium: Annagelb and Annagrün.

Recognising an opportunity to capitalise on a new trend, the Austrian–Hungarian monarchy assigned a young chemist named Adolf Patera to discover the most economical way to produce uranium glass.[29] Patera worked for a glassworks in Teplice, and conducted experiments on the effects of various elements on glass colouration. He successfully developed uranium yellow dye for glass and china in 1852.[30]

The following year, in Joachimsthal, the government opened a factory, K.K. Urangelfabrik, aka the Uranium Dye Factory,[31] housed in a building that had originally been a silver smelting manufacturing works owned by the Schlick family.[32] It soon began production of six to eight varieties of yellow dyes, including light greenish yellow, one orange and one black dye.[33]

Joachimsthal had one great advantage over other glass manufacturing areas: they had a ready source of uranium from the massive amounts of pitchblende that had already been found – and discarded as worthless – in the town's mines. With the success of the dye factory and the corresponding need for even more uranium, the derelict mines in the town were reopened and Joachimsthal experienced a revival of fortunes. By 1897 they were the world's largest uranium colour factory, making more than 12,000 kilograms per year.

In France, the glassworks Baccarat, based in Choisy-le-Roi, was known for its innovative designs and techniques, including uranium glass, which was sold under the names cristal dichroide and chrysoprase, an opaque apple-green colour.[34] Their technical advisor was Eugène-Melchior Péligot, and his knowledge of chemistry and materials science helped the company improve its recipes, contributing to its reputation for producing high-quality glass.

In the US in the late 1880s, La Belle Glass Company developed what became known as Ivory or Custard glass by increasing the concentration of uranium oxide, which made the effect more opaque. Heat-sensitive chemicals, such as gold, were added to the mix, which, when reheated during the manufacturing process, resulted in a shading effect that ranged from clear yellow to milky white at the edges.[35] Meanwhile, Burmese glass was developed by the Mount Washington Glass company. The recipe included white sand, lead oxide, purified potash, niter, bicarbonate of soda, fluor-spar, feldspar, uranium oxide and colloidal gold.[36] This formula produced an opaque glass that came in different shades, from pink to yellow. It is thought that its name was bestowed after Queen Victoria remarked that it reminded her of a Burmese sunset.

And while there were many different shades, it was the yellowish-green effect that became the most popular choice among buyers. Much later it became popularly known as Vaseline glass, due to its supposed resemblance to the famous brand of petroleum jelly.[37] There were also plenty of other companies who were using uranium to colour their glass at this time. The various producers were vying with each other to produce new colours, effects and transitions among an atmosphere of commercial secrecy.[38]

However, one of the strangest uses for this colouring was noted in 1847, when *Scientific American* reported that uranium, along with platina, titanium and cobalt, had a secondary application as a colouring agent for artificial teeth made from feldspar and quartz. By incorporating uranium as a final step in the glass-making process, just before being fired, the teeth were given an orange-yellow hue.[39] While it sounds a bit strange that this was the desired effect, throughout history dentures and false teeth had been made with ivory, gold, silver, mother of pearl or enamelled copper.[40] It was only later in the nineteenth century, mainly with the introduction of porcelain teeth, that looking natural or realistic was a desirable quality in artificial teeth.[41] And, even then, the technology wasn't quite up to scratch. Artificial teeth looked unnatural, and it was only due to a robust social contract of pretence that allowed the wearer to remain in blissful ignorance about their appearance.

Uranium oxide, along with the salts of other metallic substances, was also considered a potentially important weapon against disease and illness. This theory had a long history, dating back to the time of Paracelsus, who used toxic minerals and metals in his treatments. Considered the founder of the discipline of toxicology, Paracelsus challenged the then dominant Galenic ideas of medicine, which argued that good health was derived from a balance of the four humours – blood, phlegm, yellow and black bile. If your humours were unbalanced then illness was the likely result. The treatment for such imbalances included the therapeutic methods of bloodletting, purgatives and emetics. By contrast, by the sixteenth century, for Paracelsus and those who agreed with him, a poison in the body was best cured by a similar poison. In his mind, the therapeutic use of toxic substances could be beneficially wielded – as long as the physician was in control. After all, he asked: 'What is there that is not a poison? All things are poison, and nothing is without poison. Solely the dose determines that a thing is not a poison.'[42] The principle that this established was that everything could be toxic if taken in large enough quantities. Therefore it, was entirely possible to control the dosage and prevent harmful effects.

With this theory in mind, Christian Gmelin, a professor at the University of Tübingen, Germany, further investigated the toxicology of uranium. This research was part of a chemistry treatise, published in 1824, that described the physiological effects, on both humans and animals, of the salts of eighteen different metals, including uranium.[43]

Gmelin's *Handbuch der Chemie* described the experimentation process using uranium salts, which had been obtained from pitchblende.[44] Gmelin fed the uranium salts to dogs and rabbits in different ways and dosages to study its effects in a controlled environment. Two dogs were given their doses with food, while another dog and a rabbit received larger doses through stomach tubes. Additionally, two further dogs were even given higher doses through intravenous injection. By using these various methods, Gmelin was able to draw a conclusion about the toxicity of uranium.

He determined that while uranium was a 'feeble poison' when consumed, when administered through intravenous injection the substance proved swiftly fatal.[45]

Another researcher, C. Le Conte, carried out more experiments, this time using uranium nitrate, a yellowish crystalline substance that dissolves easily in water and is made by reacting uranium oxide with nitric acid. In 1853, Le Conte reported to the Parisian Society of Biology that he had induced nephritis, a kidney disease, by giving dogs small doses of the chemical compound.[46]

With several researchers reporting that they were able to use uranium to induce particular symptoms, the idea developed that it would also be useful to treat illnesses that presented the same side effects. Nephritis, for example, is a serious complication of diabetes mellitus, and Le Conte had noted that he observed 'sugar in the urine of dogs slowly poisoned by small doses of nitrate of uranium'.[47] From studies like this developed the hope that uranium could be used to treat the disease.

One of the first known descriptions of diabetes was back in the second century, when the Greek physician Aretaeus characterised it as 'a puzzling disease'. There had been little advancement in understanding in the intervening years and it remained untreatable, with distressing side effects and the inevitability of the death of the patient. Medical advice was limited to bed rest and a strict diet which, by the nineteenth century, included commercial products like Bonthron's Diabetic Biscuits and Bread, and The G.B. Diabetes Whisky, which promised a 'sample on application'.

Samuel West, a physician at St Bartholomew's Hospital in London, gave a boost to the potential of a uranium treatment for diabetes, publishing the results of his clinical experiments using uranium in the *British Medical Journal* in 1895 and 1896. West had given eight patients a treatment schedule of uranium salts dissolved in water, which was to be drunk after meals. He started them off with a mere one or two grains of salts and then increased slowly until they were consuming up to twenty grains two or three times a day.[48] There were often dramatic effects reported, with glycosuria – glucose in the urine – practically disappearing, and many of the patients showing improvements in their symptoms. However, there were some patients in the trial that reported gastrointestinal problems, and when treatment was discontinued for all, the effects of the disease returned practically immediately.

While the results of these tests were inconclusive, uranium treatments continued to be used in medicine and for a wide variety of ailments. According to a Dr Cook from Buffalo, uranium was great for treating urinary incontinence. An unnamed doctor claimed he had used it to cure a stomach ulcer in 1880.[49] One reported its success in haemorrhage control while another for the treatment of consumptives.[50] The pharmaceutical periodical *Chemist and Druggist* carried a recipe for a snuff, a kind of smokeless tobacco, containing one grain of uranium acetate and coffee, which was 'the latest cure for cold in the head'.[51]

A more conventional form of medication was produced by the pharmaceutical company Burroughs and Wellcome, in the form of tabloids of uranium nitrate.[52] Or by Oppenheimer, Sons & Co of London who sold palatinoids containing two and a half grains of uranium nitrate. These palatinoids were marketed as combining the benefits of uranium, which 'has been recently recommended in the treatment of diabetes by Dr S West' without the 'repugnant flavour' of his treatment.[53] Their tablets could be swallowed whole or crushed into a handkerchief and inhaled.[54]

And if you think all of that is pretty strange then step into the fascinating world of medicated wine and discover Vin Urané Pesqui! Each 24-fluid-ounce bottle of this uranium wine came with a book called *Diabetes and Its Cure by Vin Urané Pesqui*.

The book described the beverage as a powerful elixir that could instantly quench thirst, restore strength and improve bodily functions. Breathing difficulties, fatigue and lassitude were also said to be alleviated. It was even claimed that patients who drank Vin Urané Pesqui experienced a significant improvement in their appearance and temperament.[55]

The dosage – which was outlined on the label – was suggested as: 'Three small sherry-glassfuls per day, with or without water, 5 minutes before, or immediately after meals, and at night before bed time.'[56]

Uranium also found a practical application through photography, which was a relatively new field at the time. Indeed, uranium and photography were practically contemporaries. It was only a few years before Péligot had isolated the element in 1841 that the first two practical methods of permanently recording images had been developed.

In France, Louis-Jacques-Mandé Daguerre invented the daguerreotype, while in Britain, William Henry Fox Talbot developed the calotype. Both methods utilised sunlight to capture images. Daguerre made plates out of tin and later glass, while Fox Talbot used paper.

In the early days of photography, it was largely a scientific curiosity practised by wealthy men. The real breakthrough in popularity came in 1851 with the introduction of glass negatives, which allowed for multiple copies of the same image to be easily produced. This led to a fashion for cartes-de-visite, photographs mounted on a piece of card, which became a form of social currency. As the craze intensified, the number of portrait studios also correspondingly increased. London, for instance, had about twenty studios before 1853, but by the late 1860s this had increased to 284.[57]

The main technological issue with photographs at the time was the problem of fading, and the Photographic Society of London set up the Fading Committee to address the issue.[58] For some the most promising solution to the fading problem was provided by using uranium in the development process.

Charles John Burnett was a Scottish photographer who is credited with inventing the uranotype printing process in the 1850s. The uranotype used a salt, such as uranyl nitrate, to sensitise paper or other materials to light. The paper was developed with a reducing agent to produce a brown or sepia-toned image. This technique was popular in the late-nineteenth and early-twentieth centuries, particularly in scientific and technical applications.

Another photographic process that used uranium was the Wothlytype, which had been pioneered by Jacob Wothly. The Wothlytype wasn't a single idea but, as we can tell from subsequent patents, was instead a specific process, determining the way the paper was prepared and the use of uranium as a photochemical compound. The variety of tones achievable was limited, but salts such as uranium nitrate added to silver nitrate produced tones ranging from warm black to chocolate brown and brick red.[59]

Wothly's method was greeted with great enthusiasm and a bidding war to secure the patent. The patent eventually sold in France, Australia, Spain, Portugal and the United States. In Scotland permission to use the process was sold to John Urie, a photographer at 33 Buchanan Street in Glasgow. In November 1864 an advert was placed by Urie's photographic studio informing the public that they had: 'just obtained the patent rights to produce these wonderful Wothlytype Portraits so highly commended by the London Times. By the aid of this important discovery, Mr. Urie will be enabled to produce these exquisitely beautiful portraits at one-half the usual price.'[60]

In England the patent was granted to Archibald H.P. Stuart Wortley and William Warren Vernon through the company the United Association of Photography at a cost of 25,000 francs.[61] To make a return on their investment they offered licences – at a cost of ten guineas each – to photographers to use the process. Wortley had a gift for self-promotion and an impressive family connection – his sister was the goddaughter of the Duchess of Kent, Queen Victoria's mother. It was probably this family connection that allowed Wortley to proclaim: 'Her Majesty had expressed great interest in the new printing-process known as the "Wothlytype", and a portfolio of choice specimens has been, by command, forwarded to Balmoral for her inspection.'[62]

And, in an entry from her personal journal of 1865, Victoria herself wrote that she had seen 'some beautiful new kinds of Photos: called Wothleytypes [sic], printed in carbon, and which are supposed to be permanent, besides being so beautifully smooth'.[63]

Between the interest shown by Queen Victoria and positive reviews in trade and popular publications, the Wothlytype had a promising start.[64] Unfortunately the process did not gain a hold with photographers amid the eventual decline of the *cartes* market, rumours of fading, professional jealously and a flawed business model:

> The truth is that M. Wothly was an excellent operator and a good man of business. He produced first class negatives and made exquisite prints from them. He sold his patent; but he could not sell the skill to which and not to the patent the production of his capital 'specimens' was due.[65]

In August 1867 a fire broke out on the premises of the United Association of Photography, which was at 213 Regent Street, in London. It destroyed the basement laboratory and staircase to the first floor. The source of the fire was uncertain, but it was suspected that it was caused by a bursting bottle of ether or a spirit lamp that had been left burning under a vessel containing a preparation of uranium.[66] The disaster, combined with a lack of sales, led to the company's voluntary liquidation in the same year.[67]

Uranium continued to be used – as one of many chemicals – in photography, but never with quite the same enthusiasm as before. It did, however, find use in scientific experimentation, in particular one known as 'Gassiot's Cascade', or the 'electric fountain'. The cascade was described by John Peter Gassiot in 1854 in *Philosophical Magazine*. Later, the scientist and inventor John Henry Pepper offered his readers careful instructions on how to reproduce the effect so that 'a continuous series of streams of electric light seem to overflow the goblet all round the edge, and it stands then the very embodiment of the brimming cup of fire, and emblematical of the dangers of the wine-cup'.[68]

The cascade was later modified to use uranium glass, which utilises the fluorescence of the material to create an even more stunning effect, as the glass glows green. Indeed, the study of fluorescence via experiments with uranium glass was an important part of science at the time.

Gassiot's Cascade was part of a series of experiments conducted to explore the puzzling and remarkable phenomenon that had been observed while studying the effects of passing electricity through gases. This type of investigation had a long history, dating back to the eighteenth century, when it was discovered that a discharge of static electricity could be accompanied by a flash of light.[69] By the mid-nineteenth century, technology had advanced to the point where scientists could investigate the discharge of electricity through gases confined within sealed glass vacuum tubes.

Michael Faraday, an English scientist, had hypothesised that matter could exist in a fourth state, which he called radiant matter.[70] In 1838 he passed electricity through low-density gases and observed that a luminous glow developed in the discharge tubes where the electric current flowed. Faraday suggested that this glow was caused by electromagnetic radiation acting upon the gases sealed up in the tubes, much like how ultraviolet light acts on mineral crystals.[71]

Faraday lacked the technology to thoroughly test his hypothesis,[72] but his work sparked interest from other experimenters.[73] Among these was Heinrich Geissler, a German instrument maker who specialised in making blown-glass apparatus for scientific experiments. Geissler developed a new type of glass tube – sealed at both ends – that was used to enclose different luminescent gases. When electricity from a coil was passed through these tubes, they glowed brightly like an early neon sign, producing a variety of colours depending on the composition of the gases.

Geissler's tubes were fitted with two metal electrodes, which conducted electric current through the gas from one end to the other. One was called the cathode, and the other the anode. A vacuum pump was used to remove the gas, which usually was air. This resulted in a low internal air pressure, and when a high-voltage electric current passed between the electrodes, the remaining air in the tube began to glow faintly blue.[74] Even more interestingly, the rays were still observed to be present even when the more efficient vacuums had reduced the number of gas molecules inside. When the tube was coated with fluorescent material, or if the beam was pointed at a fluorescent screen, the rays became visible.[75] German physicists gave them the name *Kathodenstrahlen*, which means cathode rays.[76]

In 1869, English scientist William Crookes was experimenting with a high-voltage induction coil to create discharge. He also added a paddle wheel, which cast a shadow on the glass wall of the tube. The shadow allowed him to determine the path of the cathode rays, which were emitted from the cathode and travelled towards the anode. This led him to discover that cathode rays were actually streams of negatively charged particles, which he named 'radiant matter' and later became known as 'electrons'.

Scientists tried to change the path of these rays by using magnets to deflect them. They found that they could pass through metal foils. In 1892, German physicist Heinrich Hertz reported that the cathode rays could penetrate thin metal, gold and aluminium foil windows at the opposite end of the tube.[77] Hertz's pupil Philipp Lenard continued these experiments, replacing a portion of the vacuum tubes' glass with a piece of aluminium foil and found that this allowed some of the cathode rays to escape the tube.

The German physicist Wilhelm Röntgen became interested in these experiments and wrote to Lenard for more information.

After having had a detailed response, Röntgen replicated them and then set about exploring his own ideas. He covered his tube, which had all of its air removed, with black cardboard to prevent any fluorescence inside the tube from interfering with his observations of what was happening outside. He then placed a cardboard screen coated with barium platinocyanide nearby, which was very sensitive to light. Finally, he turned off all the lights in the laboratory. Every time he turned on the electrical current, he noticed that the screen glowed. There were various signs that this was not because of any stray cathode rays, most notably the distance they travelled – which was up to two metres. More experiments showed that these new rays could not be deflected by magnets and could pass through paper, copper and certain metals, glass and bones, producing shadows. Röntgen also discovered that if he replaced the screen with photographic plates, it was possible to develop an image that showed the different absorptions of the different materials in the image. So, materials that were of a low density were practically invisible, but denser substances, like bone, were much more clearly seen.

It was evident that Röntgen's rays were new. And in November 1895 he clarified: 'For brevity's sake I shall use the expression "rays"; and to distinguish them from others of this name I shall call them X-rays.'[78] Röntgen published a paper outlining his discovery, and in January 1896 reprints of this article, including copies of a radiograph of his wife Anna Bertha Ludwig's hand, were sent to prominent scientists in Europe. The news was picked up by many publications, including the *Neue Freie* in Vienna, *Chronicle* of London, *New York Sun* and *New York Times*.

Inevitably, scientists were the first to express their excitement. And due to the ease of replication, many took up the challenge to experiment with the new technology. But interest in X-rays extended beyond the scientific community, particularly in their potential use for photography. Both *Cassell's Magazine* and the *Clarion* published articles on the new photography in 1896, with *Cassell's Saturday Journal* reporting the following year that 'almost every week brings forth a new practical application of the Rontgen Rays'.[79]

X-rays also quickly found medical applications, with their diagnostic and therapeutic potentials being tested in hospitals. A.A. Campbell Swinton took the first clinical X-ray photograph, or skiagram, in early January 1896.[80] In July 1897, Dr William J. Morton achieved a significant milestone by taking the first whole-body radiograph of a living person in a single exposure.[81] It's a remarkable image, with some of the items worn by the patient visible, including a hatpin, necklace, rings, high button boots with nailed on heels, and a whalebone corset.

Morton also authored a well-regarded book titled *The X-Ray or Photography of the Invisible and Its Value in Surgery*. Conflicts in Europe soon highlighted the potential of X-rays in diagnosing fractures, broken bones and foreign bodies, which were commonly found on the battlefield. This led to an increased interest in X-ray technology among European military forces. Similarly, the United States army found X-rays to be useful in treating injured soldiers during the war with Spain in 1898.

At first, the technology was unreliable and required long exposure times of at least 30 minutes. Patients had to hold the film cassettes against the part of their body being photographed.[82] However, despite these limitations, by the year 1900 most major urban hospitals in the United States had already acquired X-ray machines.[83]

The technology continuously adapted and improved, and X-rays were also used for customs inspections at checkpoints and railway stations, where passengers and luggage were exposed.

But it still wasn't clear exactly what they actually were. Several theories sprang up almost simultaneously, including that the Röntgen rays were simply transverse waves or particles or sound waves or even pulses. The work on the subject was so intensive that new findings were reported almost on a weekly basis.

In January 1896, the scientist and mathematician Henri Poincaré had reported Röntgen's experiment at the weekly meeting of the French Academy of Sciences. He finished off the report with a little speculation – asking the question whether Röntgen's penetrating radiation could be found in any naturally fluorescent or phosphorescent substances.[84]

Henri Becquerel, who was present that day in the audience, became intrigued with the connection between X-rays and fluorescence and decided to investigate further. He was well-suited for the task, as he was the third physicist in a line of four who had researched the properties of both fluorescence and phosphorescence. The previous two were Antoine César and Edmond, his grandfather and father respectively.

As the chair of applied physics at the National Museum of Natural History in Paris, Henri had a vast collection of luminescent minerals at his disposal. These materials, when exposed to sunlight, absorb it and then emit light of different wavelengths from the original source. If the luminescence disappears once the light source is removed, the mineral is considered fluorescent; if it continues, it is classified as phosphorescent.

Intrigued by what he had heard that day at the Academy, Becquerel developed a hypothesis that any substance capable of emitting the radiation observed by Röntgen must possess luminescent properties. Drawing from his previous research on the phosphorescence of uranium compounds, he selected various other samples from his collection for testing. Henri suspected that some of these materials might emit penetrating radiations similar to Röntgen's X-rays.

In a methodical manner, Becquerel covered unexposed photographic plates with heavy black lightproof paper to prevent fogging by sunlight. He then placed the mineral sample on top of the paper and exposed it to sunlight for several hours by putting it on the windowsill. Depending on the specific sample, the mineral would either fluoresce or phosphoresce, both of which would have an effect on the photographic plate.

During one of his initial attempts, Becquerel exposed minerals containing uranium in this way. After developing the plate, he discovered that the radiation had indeed penetrated the paper, confirming, as Poincaré had originally suggested, that luminescent minerals emit penetrating radiation. He made several attempts and identified only one substance that emitted such radiation: a double sulphate of uranium and potassium called potassium uranyl sulphate.

At the meeting of the Academy of Science in February 1896, Becquerel presented his findings. He reported that various materials emitted rays capable of penetrating the thick black paper and exposing photographic plates, even if only faintly. His claim was fairly unremarkable; he simply asserted that the radiation was strong enough to darken the plates.[85]

Becquerel worked to refine his results and confirmed that any material containing uranium caused the plates to darken images to be imprinted on the photographic paper, regardless of whether they exhibited phosphorescence or not. While this should have cast doubt on Becquerel's theory that the radiation was related to phosphorescence, he remained committed to this idea.

On 26 and 27 February, Becquerel prepared sets of potassium uranium sulfate crystals and photographic plates, intending to expose them to sunlight by placing them on a windowsill as before. However, inclement weather prevented him from carrying out this plan, and he postponed, placing the samples in a drawer to wait for better weather.

When Becquerel retrieved the samples from the drawer a few days later and developed the photographic plates, he was surprised to find imprinted images. He had expected to find faint images, but the uranium had affected the plates without exposure to sunlight. This discovery contradicted the notion that sunlight activated the minerals and suggested that the connection between phosphorescence and X-rays would have to be abandoned. Upon further investigation, Becquerel found that the emission of the penetrating radiation was not affected by changes in temperature or chemical reactions. This led him to conclude that the uranium itself emitted rays, representing an entirely new property of the material.

His discovery of a radiation emitted by uranium was published in *Comptes rendus de l'Académie des Sciences* (*the Proceedings of the Academy of Sciences*). While he received a great deal of acclaim for this discovery, including a shared Nobel Prize for Physics in 1903, Becquerel played down his contribution, claiming that it was more of a family affair: 'These discoveries are only the lineal descendants of those of my father and grandfather on phosphorescence and without them my own discoveries would have been impossible.'[86]

And indeed, this muted response to the discovery of these '*rayon uraniques*' or 'Becquerel rays', as they were known at the time, was rather typical of their overall reception. Surprisingly little work was done on them over the next few years. Even Becquerel himself wasn't that interested. He published seven papers on the phenomenon in 1896 and two in 1897, before returning to his other topics of study.[87]

But while there wasn't much interest in Becquerel's rays, the popularity of X-rays was going from strength to strength. Beyond the world of medicine and photography, there were some entrepreneurs who recognised X-rays' potential for entertainment and profit. Public exhibitions of X-ray photographs were a popular attraction in many cities, and traveling X-ray shows emerged, offering curious viewers the opportunity to see the inside of their bodies.

At the forefront of this style of scientific entertainment was the inventor Thomas Alva Edison, whose team at Menlo Park developed an enhancement to Röntgen's original technology and promptly licensed it to a local manufacturer. The device was later introduced as the Thomas A. Edison X-ray Kit – a handheld device where the image was projected onto a screen, allowing viewers to get a glimpse of their own bones.

There were also other demonstrations of X-rays, such as coin-operated technologies which were likened to a 'match or chocolate automatic machine'.[88] These devices were glass cases that housed an X-ray tube and a high-voltage coil, with a slot for inserting a hand or an object, like a purse containing coins. By looking through a double eyepiece on top, one could see the bones or the coins inside. These were marketed as a way for public places like restaurants and bars to generate money, with adverts promising: 'The latest application of its remarkable power is the X-ray slot machine. A nickel and you look through anything. Increases cigar, drug and other trade 50 per cent. Big money exhibiting it.'[89]

People were fascinated by the ability of X-rays to reveal the hidden and often mysterious aspects of the human body. However, this new form of entertainment was not without risks, as many of the early X-ray machines were poorly regulated, and some operators used excessively high doses of radiation, causing burns, radiation sickness and even death. Later, as the dangers of over-exposure to X-rays became more apparent, those X-ray slot machines instead dispensed a little card with an illustration of a skeletal male or female with a hat or a pipe or a bonnet.

There was also unease about the moral implications and the potential indecency they posed, especially around the fears of being able to see through clothes. A concern, expressed in rhyme in an 1896 edition of *Electrical Review*.

The Röntgen rays, the Röntgen rays –
What is this craze?
The town's ablaze
With this new phase
Of X rays ways.

I'm full of daze,
Shock and amaze;
For nowadays
I hear they'll gaze
Thro' cloak and gown – and even stays!
These naughty, naughty Röntgen Rays![90]

This potentially rather saucy side effect of the new technology was also expressed visually in a short silent comedy movie directed by George Albert Smith and released in 1897. The film, which is called *The X-Rays* or *The X-Ray Fiend*, shows a couple, played by Laura Bayley, who was the wife of the director, and an English comedian named Tom Green. While sitting on a bench the flirting pair are irradiated by a man carrying a movie-camera shaped device marked 'X-Rays', which allows the viewer to see through their clothes. The pair are now visible as skeletons, and in a lovely touch the umbrella that Bayley is holding also loses its fabric panel and only the handle and ribs are evident.[91]

The idea that X-rays somehow reveal morality continued when the term was co-opted to mean anything slightly transparent or titillating. One of the best examples of this was the 'X-ray' skirt – a new fashion in women's clothing that was the subject of several horrified press reports during 1913. Mrs Almena McDonald, President of the Woman's Christian Temperance Union, made her thoughts on the matter quite clear, stating that the X-ray skirt was not just a matter of fashion but one of morality. As far as she was concerned that particular style of dress was one step up from young girls parading the streets 'wearing nothing at all'.[92]

However, rather than a harbinger of moral decay, some hard-hitting investigative journalism concluded that the issue was a perfect storm of the normal practice of wearing dark-coloured stockings, the fashion for wearing white gauzy skirts that year and bright sunlight, which meant the wearer's legs became visible through their outer layer. The moral outrage over X-ray skirts cleared up later that year and was barely mentioned again.

Uranic rays didn't have a hope of comparing with their more glamourous relative, and by the turn of the century uranium was still being labelled by the *Reader's Digest* as 'a worthless metal not found in the United States'.[93]

They were wrong.

URANIUM, VANADIUM AND RADIUM

2

Marie Skłodowska Curie was an outlier who had been intrigued by Becquerel's findings when she read about them in *Comptes rendus*. She was particularly interested in investigating whether uranium was truly the only element that emitted these types of rays.

So, she took over a disused storeroom at the École de Physique et Chimie Industrielles (EPCI) in Paris and started her PhD research by measuring the radiation emitted by uranium compounds, before expanding her investigations to other elements. Initially, the focus was on establishing an accurate method of measuring the intensity of the rays. Although Becquerel's experiments with photographic plates had produced some striking images, there was no effective way of quantifying the radiation. In the end she decided to use a piezoelectric electroscope that her husband, Pierre, and his brother, Paul-Jacques, had developed back in 1880. With this instrument, which was incredibly sensitive to even the faintest trace of current, she was able to detect the electrical effects in the air caused by the radiation, which would later be identified as part of the process of radioactive decay.

As we learn from Skłodowska Curie's laboratory notebooks of December 1897, and from subsequent reports, she tested every mineral and rock that she could get hold of, which included the collection at the EPCI, but she also borrowed samples from mineralogists and chemists. She may have tested practically every known element, but only two – uranium and thorium – seemed to give off this energy and acted in the same way.

These initial tests led to her pinpointing further research on minerals containing uranium, particularly pitchblende, which is composed of approximately 60 per cent of the element. Based on Becquerel's findings that the intensity of radiation emitted by a uranium compound is proportional to its uranium content, she expected that 1 gram of pitchblende would have the same effect on her measuring instrument as 0.6 grams of uranium. However, she was surprised to discover that the electrical energy emitted was actually several times greater than that. She repeated the experiment many times to make sure there wasn't an error. This led her to conclude that 'these minerals may contain an element much more active than uranium'.[1] She told her sister Bronya: 'The element is there and I've got to find it.'[2]

To achieve this goal, now working with Pierre, she used the painstaking technique of fractional crystallisation to isolate the unknown substance from the complex and impure pitchblende. This involved various chemical procedures to separate it into distinct parts, which were then tested using the electroscope. The Curies identified that certain separated compounds, obtained in incredibly minute amounts, demonstrated strong radioactivity, leading them to conclude that they had discovered trace amounts of a new substance, since there was no existing explanation for this phenomenon.

They hypothesised that there were two new elements: polonium, named after Marie's birth country, and radium, named after the Latin word for ray, 'radius'. Over the course of 1898, having announced their preliminary results at the regular Monday afternoon meetings of the Academy of Science, they confirmed the existence of these two new elements. It was also during this time, in a paper of July 1898, that Skłodowska Curie first used the term radioactivity to describe what had previously been known as uranic rays.

To further their research, the Curies required more pitchblende and approached the Austrian Academy of Sciences in Vienna for assistance. The academy readily provided them with ample amounts of the material, from the Joachimsthal mines which were now under their control, as it was considered worthless after the initial extraction of uranium.

After discovering that polonium was difficult to separate, Skłodowska Curie shifted her focus to radium, which appeared to be more powerful and easier to extract. However, the process of producing radium samples proved to be lengthier than expected. It was back-breaking work, with Marie later recalling:

> Sometimes I had to spend a whole day mixing a boiling mass with a heavy iron rod nearly as large as myself. I would be broken with fatigue at the day's end. Other days, on the contrary, the work would be a most minute and delicate fractional crystallisation, in the efforts to concentrate radium.[3]

To expedite the process, the Curies collaborated with the chemical manufacturing company Société Centrale des Produits Chimiques, which assisted in identifying successful extraction procedures. While the company handled the strenuous task of heating and boiling large vats of minerals, the scientific analysis was performed at the Curies' laboratory. Over the next four years they processed eight tons of pitchblende residues before announcing, on 21 July 1902, that they had isolated one-tenth of a gram of the compound radium chloride.

While the Curies were working on isolating enough radium to establish its atomic weight, an essential step for it to be officially classified as a new element, there were also plenty of other scientists working on this new discovery. Ernest Rutherford, a physicist from New Zealand, embarked on a mission to understand radioactivity, beginning his research at the Cavendish Laboratory at the University of Cambridge in late 1897, and continuing at McGill University, Montreal, Canada.

Through these experiments his team demonstrated that radioactive substances emitted not one, but two distinct types of rays. According to legend, Rutherford didn't know what to call them so simply went with the first two letters of the Greek alphabet – alpha (α) and beta (β). The names stuck.

They conducted experiments to confirm the absorption capabilities of the various radiations. He covered the uranium material with aluminium sheets of varying thickness and used an electrometer to measure the electric current generated by the transmitted radiation.[4] It was determined that the alpha rays moved with a velocity of 20,000 miles per second but could be stopped easily by a single sheet of paper, as it was cut off or absorbed by it. The second type of radiation, beta, had rays that were streams of electrons and had a hundred times more penetrating power than alpha particles. They could pass through thin filters of aluminium but were stopped by sheets of lead or other heavy metals.

Over the next few years many researchers carried out experiments to learn more about these different radiations. Physical chemist Paul Villard discovered another emission from radioactive materials, which he named gamma (γ) rays. Unlike alpha and beta rays, gamma rays are highly penetrating and travel at the speed of light. Initially thought to be without charge, they were later shown to be a form of electromagnetic radiation similar to X-rays.

Between 1900 and 1902, Rutherford and chemist Frederick Soddy collaborated on nine papers while at McGill University.

They proposed their theory that atoms were breaking apart and forming new types of matter, and that the unusual behaviours of known radioactive elements were due to their perpetual break-down and a series of successive transformations into something else as they decayed.

The scientific community gradually developed a deeper understanding of radioactivity over a period of several years. It was discovered that naturally occurring radioactive substances could be classified into three distinct groups: the 'thorium', the 'actinium' and the 'uranium or radium' series. For example, ura-nium undergoes a series of transformations to become radium, followed by radon and eventually polonium. This process con-tinues until a stable element, such as lead, is reached, which no longer emits radiation. Each stage in the process has a specific 'half-life', which is the average amount of time it takes for half of the original atoms to change or decay into a new element.

In 1902, Soddy left Montreal and spent a year in London working with William Ramsay, a professor of chemistry at University College London. During this time, they conducted experiments and were able to provide measurable evidence for the transformation of radioactive elements. Later, working with Ruth Pirret, Soddy discovered that uranium existed in nature as a mixture of two isotopes:[5] 'We are therefore faced with the possibility that uranium may be a mixture of two elements of atomic weights 238.5 and 234.5 which ... are chemically so alike that they cannot be separated.'[6]

This reflected a problem that chemists were having. Researching the disintegration of elements meant investigating a whole new series of spontaneously formed elements. These had similar chemical properties, but different masses and decay rates. They could not be separated through chemical means because of their identical chemical properties, but their atomic masses, the rates they decayed, and radiation types differed.[7] As

Soddy said in his address after receiving the 1921 Nobel Prize in Chemistry, it was like having atoms with 'identical outsides but different insides'.[8]

The need to come up with a new term was driven by Soddy, after he apparently grew tired of writing out the phrase 'elements chemically identical and non-separable by chemical methods'. And there would have been many times it needed to be written: by 1910, 44 of these new elements had been identified.

The term 'isotope' was officially first used by Soddy in a short letter to the editor published in the December 1913 issue of *Nature*. While it was a neat solution to the problem, Soddy failed to give credit to the person who came up with the name.

Earlier that year Soddy attended a dinner party hosted by industrial chemist Sir George Beilby, where the conversation turned to some of the astonishing recent developments in radiochemistry. Another guest as the party, Dr Margaret Georgina Todd, who had a mastery of the Greek language, suggested using the word isotope as a way of solving Soddy's problems. (In Greek, 'iso' means same and 'topos' means place.)[9]

While researchers were investigating atomic structures, medical practitioners were interested in finding potential uses for radioactive elements. Learning from the recent innovation with X-rays, attempts were made to produce medical photographs, but the radiograms created by radium rays were fuzzy and imprecise, leading to their discontinuation.

X-rays had also been used as treatments. In early 1896, Emil Herman Grubbé, a medical student and owner of an electric light bulb company, was among the first to suggest that X-rays might have therapeutic potential beyond their diagnostic use. Grubbé and a colleague noticed symptoms of acute dermatitis on their hands after X-ray tests, leading a medical professor to suggest that the rays may be effective in treating diseased tissue. This theory was tested on patients, including Rose Lee, who had advanced breast cancer, and she underwent multiple X-ray treatments before dying a month later. While the treatment had not been effective, and arguably had been delivered too late for Lee, experiments into its therapeutic use continued.

In 1900, the German dentist Otto Walkhoff, who had been a pioneer of X-rays in his work, conducted the first tests of the effects of radium salts directly applied to human tissue. Walkhoff exposed his arm to radium rays in two twenty-minute bursts, noting that his skin became very inflamed in a comparable reaction to the kind that resulted from overexposure to X-rays. Pierre Curie and Henri Becquerel also reported their experiences of tissue damage resulting from radium exposure. Intrigued by these results, Dr Henri-Alexandre Danlos conducted successful experiments using radium salts to treat skin conditions. The similarity between X-ray burns and radium burns drew analogies between those treatments, leading many X-ray pioneers to consider radium radiation as an alternative treatment method. Curietherapy, or radium therapy, began to develop rapidly among a handful of private practitioners throughout the world.

The apparent medical advancement of radium sparked a period of 'radiomania' similar to the X-ray mania of only a few years earlier. The radioactive properties of radium were seen as beneficial, with people believing that exposure to it could boost energy levels and improve overall health. As a result, radium-based tonics, foods and cosmetics became widely popular. The word 'radium' was exploited by advertisers and commercial businesses to signify scientific progress and technological advancements, even in products that had no real association with the substance.

As the worldwide fascination with radium grew, there were concerns that the hard-to-produce substance would become even scarcer due to demand. To counter this the Austrian Academy of Sciences put an embargo on the export of pitchblende residues from Joachimsthal in late 1903.[10]

While the increased interest in pitchblende had benefited the town financially, they also profited from the widespread belief in the health benefits of radium. The early 1900s saw the establishment of the Experimental Spa Institute and a number of other similar sites where people could bathe in water rich with radioactivity and inhale radon gas. Additionally, spa hotels like the Radium Palace Hotel were founded.[11] During its peak years, the Radium Palace Hotel welcomed approximately 2,500 guests annually, including J. Robert Oppenheimer, a young physicist who would later head the Los Alamos Laboratory, where radioactive elements had a very different purpose.[12]

The increase in price and embargo of radium had made it challenging for scientists and medical professionals to continue their experiments with it. As a result, there was a search for alternative radioactive elements.

In Finland, monazite – an ore containing thorium – deposits were discovered in Veaborg and Uleaborg. Uranium-bearing torbernite was found near Lake Onega in northern Russia. More uranium ores were found in Australia and in Cornwall, where Uranium Mines Ltd operated the South Terrace Mine, five miles from St Austell.

Under the mine manager, Captain W.R. Thomas, the company set up works in London to extract the metal from the ore. It was then roasted, crushed to a fine powder, dissolved in acid, precipitated with chemicals, filtered, dried and pounded into a yellow powder. They hoped that the powder would be used in steel after research carried out by major steel and gun manufacturers in England and Germany had revealed that adding a small percentage of uranium to steel could increase its elasticity and hardness. As a result, it became highly desirable for making guns, armour plates and other such products.

However, Uranium Mines Ltd was soon voluntarily liqui-
dated, with shareholders questioning the cost of the project,
the high pay of the directors and the lack of dividends. In
March 1891, a stormy meeting was held in which the chair,
Sir Alexander Armstrong, reported on the status of the company. A few weeks before the meeting, a shareholder had
taken Uranium Mines Ltd to court for selling shares under
false pretences, as the prospectus claimed that the company
had found a uranium lode recently, but they were referring to
the one found many years previously. The judge ruled that this
was essentially false selling and designed to deceive shareholders. While Uranium Mines Ltd had operated fraudulently,
other uranium processing companies were also struggling,
as it proved difficult to extract the ore, making all uses –
including the much-lauded uranium steel – too costly to be
sustainable.

In the United States uranium-bearing ore was found in a
number of small mines over what is now known as the Four
Corners region – where New Mexico, Arizona, Utah and
Colorado meet.[13] The area is largely comprised of the land of
Diné Bikéyah – also known as the Navajo Nation – with four
sacred mountains, Tsisnaajinii (colonial name Blanca Peak) to
the east, Tsoodził (Mount Taylor) to the south, Dook'o'oosłííd
(San Francisco Peak) to the west, and Dibe' Ntsaa (Mount
Hesperus) to the north.[14]

It is a terrain with a long and tragic history, and it is where
the Diné – the Navajo people of the American south-west – were
forcibly removed by the US government.[15] They were driven
from their homeland to the inhospitable and barren land of
Bosque Redondo, in New Mexico.[16]

Decimated by disease and poverty, they were later allowed
to return to a small area of their homeland after the signing
of the Navajo Peace Treaty in June 1868. The US government

established a precedent of unfulfilled commitments, unequal treatment, and a failure to comprehend or address the requirements of Indigenous Peoples. And this already destructive relationship was compounded by the richness of the land itself, which was in turn exploited for a rapid succession of energy resources – from gold to coal, and later uranium, first for its radium content and then for its own atomic properties.[17]

In the region pitchblende was widely scattered in hard granite rock but there was also plenty of carnotite, a bright greenish-yellow mineral, found in sandstone and in petrified wood. Carnotite was first formally identified in 1899 and was named by mineralogists Charles Friedel and Eduard Cumenge after the French chemical engineer Adolphe Carnot.[18] And it was confirmed that it was made up of a lower concentration of uranium but also vanadium.[19]

Vanadium, while not radioactive, was a valuable metal. Following the Franco–Prussian War of 1870–71, there was a surge of interest in steel alloys, such as ferro-uranium and ferro-vanadium, for the European armaments industry. These alloys were discovered to be effective hardening agents when added to steel, leading to their widespread use in a variety of products, including automobile parts and gun barrels.

The largest supplier was the American Vanadium Company, established in the Andes Mountains of Peru in 1906 by Pittsburgh businessmen Joseph and James Flannery.[20] The brothers eventually became the world's largest supplier of vanadium. Their ferro-vanadium was marketed to the railroad companies for use in both locomotive construction and the rails themselves, and it was used in the construction of the Panama Canal and was prized by Henry Ford.[21]

Vanadium was vital to the development of the Ford Model T, with Ford himself claiming to have recognised its potential after examining the wreckage of a French race car which had crashed in Michigan.[22] Ford and his team of engineers saw that vanadium steel had a high tensile strength that was nearly three times greater than cheaper, lower-grade steels. They decided to incorporate it into many of the cars produced by the Ford Motor Company, especially the bits that endured a large amount of wear and tear, such as the crankshaft, forged front axle and the wheel spindles. Model N and R cars were marketed with the claim that:

> We are still the only American makers who use Vanadium steel in motor car construction AND FROM NOW ON Vanadium-chrome steel will be used throughout in all Ford models – runabouts as well as six-cylinder touring cars. Let others follow as soon as they can – we figure they're about a year and a half behind at present.[23]

While not as high quality as the Peruvian sites, there were vanadium mines scattered throughout Arizona, New Mexico, Utah and Colorado during this period, and several mill operations extracting vanadium and uranium from carnotite.[24]

Carnotite was also confirmed in 1911 to contain radium, although it proved challenging to extract. Those companies that initially tried, like the Welsh-Loftus Uranium and Rare Metals Company, found the high cost of chemicals and the low recovery rates of the radium a poor return on their investment.[25]

Despite the initial challenges, production continued to increase as the technology improved, and by March 1912, Colorado Plateau carnotite had become one of the primary sources of radium.[26] A report from the US Bureau of Mines revealed that the quantity of radium obtained from carnotite was nearly four times that from pitchblende.[27]

While the Bureau of Mines reported that the radium-containing ores found in America were among the richest deposits of radium in the world, they also expressed concern about how much was being exported.[28] In a report to Congress, the Bureau suggested: 'The uranium deposits of Colorado and Utah are being rapidly depleted for foreign exploitation. It would seem to be almost a patriotic duty to develop an industry that will retain the radium in America.'[29] There were various attempts designed to stop the valuable ores from being sent to Europe, from a 25 per cent import duty imposed on the cheaper foreign-manufactured radium and a failed attempt to nationalise all rights and ownership of radium-bearing ores found on public land.[30]

With radium selling for $90,000 a gram, there was a huge incentive for miners to flood into the Four Corners region. Between 1913 and 1930 several other radium companies became operational, and all began to acquire mining claims in the Paradox Valley or nearby.[31]

The industry grew to be dominated by the three largest companies: The Radium Company of Colorado, the US Radium Corporation, and Radium Chemical Co.[32] The latter was owned by the Flannery brothers.[33] Joseph and James had established the first company to produce radium on a commercial scale in the United States in 1911 with the founding of Standard Chemical Company. Uniquely, they became interested in radium extraction not purely for the profits, but because they had been devastated by the death of their sister, Eleanor, in 1910 of uterine cancer.[34]

The Flannery brothers were inspired by the purported success of radium treatments in Europe and vowed to produce it commercially in the United States from their existing carnotite deposits. Like many others, including respected medical professionals, they believed that radium had the best potential for a cure for cancer, and they sold their mine in Peru to focus on their operations in the US.[35] In the space of four years they purchased 90 to 100 carnotite claims and set up an extraction site 20 miles southwest of Pittsburgh.[36] They also established the second processing mill in North America in Uravan, Colorado.[37] To improve their extraction processes, they brought in experts from Europe, including Otto Brill from Joachimsthal, who made significant improvements to the methods that had previously been utilised.[38]

The Flannery brothers' extraction factory began receiving ore in 1911, and by 1913 they had produced their first purified radium, although it was a costly endeavour.[39] In May of the same year, Standard Chemical became the first American company to put radium on the market.[40]

The United States became the dominant player in the radium market, producing 80 per cent of the world's supply, most of which was either used in medicine or in self-luminous paint.[41] This was the result of the previously widely reported discovery that radium salts actually glowed in the dark, a result of radiation interacting with the nitrogen naturally present in the air. The vibration generates a surge of energy, manifesting as a shimmering light. Adding a sticky substance to these salts made a self-luminating, permanent glow-in-the-dark paint, which became a source of immense commercial interest. The market for this increased with the outbreak of the First World War in 1914, which led to a huge demand for radium-painted watches, clocks, compasses, binoculars, gun sights and aeroplane instruments.[42]

As Western countries aimed to harness their domestic natural resources, they also looked towards exploiting the regions subjected to colonial rule. This ambition was particularly evident in the case of Congo, in central Africa, towards the end of the late-nineteenth century. Like many other regions in Africa, Congo was occupied by European colonial powers, and exploited for its mineral wealth, which included copper, tin, iron and silver, as well as its plentiful natural rubber.

Initially privately controlled by King Leopold II of Belgium and renamed the État Indépendant du Congo (Congo Free State), this period of colonisation was marked by war, widespread human-rights abuses and brutality. Forced labour, torture and starvation were rampant, resulting in the deaths of an estimated 10 million Congolese.

After international condemnation, the country was annexed by the Belgian government in November 1908 who, after renaming it the Belgian Congo, controlled the area as a colony until its independence in 1960. In order to continue to exploit the land and its people, the Belgian government went on to provide financial support to the Union Minière Holding Company, one of the largest investment consortiums in Europe.[43]

In 1907, Union Minière acquired the prospecting and surveying services of the Englishman Robert Sharp. Despite being essentially a nepo hire (he was a friend of one of the bosses' families), Sharp quickly adapted to his role and received training before heading to Africa:

> It was decided that before leaving England I was to receive some intensive training in mineralogy and surveying. For the former I went to an old man who had a little shop in Regent Street. He taught me a lot about the identification of mineral specimens and geology, and I also learnt how to use a blowpipe. For survey instructions I went to the Royal Geological Society, where I got a grounding in the use of the sextant, identification of stars, map making etc.[44]

While Sharp may have lacked experience, he made up for it with his eagerness to learn and dedication to the job.

A few years later Sharp discovered a bright yellow material in the Haut-Katanga Province, an area that later became known as Shinkolobwe.[45] However, the outbreak of the First World War prevented him from capitalising on this discovery: Sharp had to leave for Europe to serve, and Union Minière had to focus on copper production for the Allied war effort.[46] As Sharp later recalled in his autobiography: 'At this time uranium had no commercial value except as the source of radium and it was not until the atomic age began to dawn that the main value of the property was realised.'[47]

After the war had ended, Sharp revisited the area and found that his original claim was still open.[48] It was eventually developed and exploited by the Union Minière, who conducted an analysis and found that the deposits were of extremely high grade, with up to 68 per cent uranium, the highest ever found at the time.[49]

Congolese labourers were employed to work in the mines under poor conditions and with low pay.[50] The highest-grade material was hand-separated at the surface before being transported to Belgium. The lower-grade ore, which was mostly torbernite, was simply dumped.[51]

After the shipment of uranium ore arrived in the port city of Antwerp, it was transported to the extraction factory in Olen.[52] There, a preliminary purification plant had been constructed in 1920 by a new partner, the Société Métallurgique de Hoboken. By 1922 the first radium bromide had been produced and was delivered to the Union Minière headquarters in Brussels. It was found that a gram of radium could be produced from ten tons of ore, compared to the 125–500 tons of American carnotite needed to produce the same amount.

Edgar Sengier, the director of Union Minière, announced the company's intention to sell their radium at almost half the price charged by outfits such as Standard Chemical and the Radium Company of Colorado.[53] The Belgian company's rich claims and exploitative labour practices allowed them to keep the price low, making it impossible for US producers to compete, and many companies went out of business.[54] By the mid-1920s Belgian sales accounted for approximately 80 per cent of the world market.[55]

However, their domination was eventually broken in 1930, when Eldorado Gold Mines discovered rich pitchblende deposits at Great Bear Lake in Canada, leading to increased competition in the radium market. In 1933, the first ore was processed in Port Hope, Ontario, and as a result the price of radium dropped to $20,000 per gram.[56] But the importance of radium in the world market was soon to be eclipsed by another radioactive element.

ENTERING THE ATOMIC AGE   3

During the 1930s, scientists faced a perplexing challenge: how to understand the inner workings of the atom. While the concept of atoms, or at least the understanding that matter had to be composed of something, had been around since ancient times, it was not until the early 1800s that chemist John Dalton provided evidence of their existence. However, the mystery of what was actually happening inside the atom continued to intrigue.

At the time the most widely accepted theory of atoms had been put forward by the English physicist J.J. Thomson, who was the head of the Cavendish Laboratory in Cambridge. After having identified the electron, a subatomic particle, Thomson suggested that the atom was made up of negative electrons spread through a sphere of positive charge. In this model, electrons were envisaged as plums dotted through a pudding – which is why it was known as the 'plum pudding model'.

This understanding allowed scientists to investigate the atom by bombarding a substance with particles, specifically positively charged alpha particles. These could either pass through the object unchanged, be absorbed by it or bounce back – a process known in physics as back scattering – and be detected by a measuring device. By analysing the amount of scattered radiation at different angles, physicists could make educated guesses about the underlying structure of the atom.[1]

Now at the University of Manchester, Ernest Rutherford designed a new experiment to investigate the interior of an atom in 1911. Carried out by his research student Ernest Marsden, the experiment called for alpha particles to be beamed at a very thin sheet of gold foil. It was expected, as determined by the plum pudding model, that the positive part of the gold would slightly repel the charged particles. These would then change course by a small amount. Instead, it became clear that while most of the alpha particles passed straight through the foil, some rebounded. This was remarkable, as Rutherford explained later: 'It was as though you had fired a fifteen-inch shell at a piece of tissue paper, and it had bounced back and hit you.'[2]

Rutherford's conclusion was that the beamed alpha particles were encountering a massive charge – but only occasionally. This suggested that the atom was mostly sparsely populated, but that something was there which was deflecting the particles. This 'something' was concentrated in a very small area, allowing most of the particles to pass through without any interference.[3]

After further experiments in conjunction with his colleague, German physicist Hans Geiger, Rutherford's scattering theory was presented at a meeting of the Manchester Literary and Philosophical Society, and in May 1911 he published a paper in the *Philosophical Magazine*.[4] These experiments had essentially overturned Thomson's plum pudding model, but it took many years of research before a full explanation was determined.

By 1914 there was a new, widely accepted theory, in which electrons move around the nucleus in random orbits like planets orbiting a star. This was further elaborated on by Danish physicist Niels Bohr.[5] The visual icon associated with this model, which shows a small nucleus and an electron circling around it in a fixed orbit, came to define the atom in the public imagination and was eventually used as the logo for the Atomic Energy Commission.

Later, in 1919, Rutherford bombarded and breached the nucleus itself, finding that it absorbed the alpha particle and transformed. Thus, a nucleus of nitrogen changed into a nucleus of oxygen and emitted what Rutherford termed a proton.

So, by the early 1920s, a lot was known about the structure of the atom: there were electrons and protons, and there was a nucleus. That those electrons and protons should be, in order for the atom to be electrically neutral, found in equal quantities was also understood. That some information was still missing was suggested by the observable fact that some atoms, such as helium, had an atomic number of two but a mass of four. The mass of the helium atom was simply too big to just consist of protons and electrons. It suggested the existence of another particle inside the nucleus.

There were researchers, like Walther Bothe, Herbert Becker, and Irène and Frédéric Joliot-Curie, who got close to the explanation. They all, like so many other physicists at the time, were bombarding various substances with alpha particles from radioactive sources and observed a new form of radiation that had great penetrating power. This was widely interpreted as being an artificial gamma ray.[6]

Rutherford was not convinced that this explanation was the correct interpretation of their experiments. Now Cavendish Professor of Physics at Cambridge, Rutherford asked physicist James Chadwick to investigate.[7] The hypothesis was that there was another particle inside the nucleus which was the source of this energy. Chadwick, who was assistant director of the laboratory, worked on this problem in early 1932 before confirming the existence of a new particle and solving the enigma.[8] It had been an intense search and Chadwick's experiments had taken ten days, during which he had slept for only three hours each night; when it was over, he said: 'Now I want to be chloroformed and put to bed for a fortnight.'[9]

Upon the discovery of this new particle – named the neutron – scientists finally achieved a complete picture of the basic structure of the atom. They understood that the nucleus is composed of tightly packed protons and neutrons, forming a compact mass. The electrons of the atom occupy the vast space surrounding the nucleus. As Rutherford himself declared, this was 'the greatest discovery since the artificial disintegration of the atom'.[10]

As such, the unearthing of the neutron had far-reaching implications beyond simply understanding the fundamental composition of the atom. It led to the initiation of several research projects, as scientists recognised the potential of this newly discovered particle. With its significant mass and neutral charge, the neutron could be used as a tool to scrutinise other atomic nuclei, which would lead to further developments in nuclear physics and radiation science.[11]

It was also proposed that neutrons and other particles could be used to create new elements, ones that didn't exist in nature. By the 1930s a total of 88 naturally occurring elements had been identified, including six since the beginning of the century. It had always been known that there were other elements that were yet to be discovered. Indeed, when Dmitri Mendeleev introduced his periodic table in 1869, there were 60 listed chemical elements and blank spaces for unknown ones.

At the forefront of this research were the scientists at the Radium Institute in Paris. Led by Irène Joliot-Curie, the eldest daughter of Pierre and Marie, and her husband Frédéric, their experiments concentrated on the lightest elements, those that have the smallest number of protons and electrons and can be found towards the top of the periodic table. They began by bombarding thin sheets of aluminium with alpha radiation from a powerful source of polonium. This action was found to actually induce radiation in the form of radioactive phosphorus.[12] Further observations revealed that the aluminium foil remained radioactive for a short time even after the alpha particle bombardment stopped, before eventually decaying into stable silicon.

The detection of this induced radioactivity was made possible with the use of a Geiger–Müller counter, a device designed by Hans Geiger in 1908. This counter identified the pulses that the alpha particles produced, which are heard as clicks. By the time Wolfgang Gentner, part of the Joliot–Curie team in Paris, built their device, the technology used in Geiger–Müller counters had improved significantly.[13]

There was no doubting what they had done: they had converted stable atoms into radioactive atoms – transforming one element into another spicier version. The creation of the first artificial radionuclide cemented their place in history and established the ability to make radioactive elements in the laboratory.

In one of those occasional neat ties of history, Marie Skłodowska Curie was able to witness this groundbreaking discovery. Frédéric recalled:

I will never forget the expression of intense joy which overtook [Mé's] face when Irène and I showed her the first [man-made] radioactive element in a little glass tube. I can see her still taking this little tube of the radioelement, already quite weak, in her radium-damaged fingers. To verify what we were telling her, she brought the Geiger Müller counter up close to it and she could hear the numerous clicks ... This was without a doubt the last great satisfaction of her life.[14]

Marie died on 4 July 1934 of aplastic anaemia – likely caused through her work with X-rays during the First World War. She and Irène revolutionised battlefield medicine with the creation of the 'radiological car', a specially adapted vehicle containing an X-ray machine, nicknamed the 'petite Curie'. Marie not only learned to drive so she could operate one of these mobile X-ray units herself, but also trained over a hundred women volunteers in the use of the equipment. By the end of the war around a million wounded soldiers had been helped.[15]

Further experiments produced radioactive nitrogen from boron. Magnesium, bombarded, became an unstable isotope of silicon. As they decayed to their stable elements, each of these radioelements had varying half-lives, ranging from fourteen minutes to less than three.[16] The results, which were published in French in *Comptes rendus* and in English in the 10 February 1934 issue of *Nature*, stimulated similar experiments in other laboratories.[17]

Irène and Frédéric Joliot-Curie were awarded the Nobel Prize for chemistry in 1935 for this work. They were only the second couple to achieve this feat – the first being Pierre Curie and Marie Skłodowska Curie in 1903. The parallels were noted by Joliot-Curie herself in her Nobel lecture when she acknowledged her contribution as part of an 'extraordinary development' in the science of radioactivity that had been started less than 40 years before by her mother, father and Henri Becquerel.[18]

Most experiments aimed at creating artificial radioelements and studying the structure of the atom relied heavily on particles from the known radioactive elements, like the polonium used in Paris, uranium and thorium. However, with advancements in atomic research, scientists realised that the energy produced by these particles was inadequate for their purposes. They needed more, with Rutherford saying that he wanted 'a copious supply of atoms and electrons which have an individual energy far transcending that of the alpha- and beta-particles from radio-active bodies.'

Rutherford knew he needed a high-electric-voltage accelerator but lacked the technical know-how to build one himself. At his laboratory in Cambridge, the task of developing the desired particle accelerator fell to John Cockcroft and Ernest Walton.[19] In the United States, Robert Van de Graaff experimented with high voltage increases. In Germany, Norwegian-born engineer Rolf Wideröe focused on a linear accelerator design which utilised a series of straight tubes.[20] Indeed, it was Wideröe's subsequent paper that caught the attention of Ernest Lawrence at the University of California, Berkeley, who went on to develop a circular particle accelerator – later dubbed the cyclotron.[21] Lawrence proposed the idea – where particles were bent into a circular path using a magnetic field, accelerated and then smashed into selected targets – but it was his graduate student Milton Stanley Livingston that turned the idea of 'a proton merry-go-round' into reality.[22]

In January 1931, the first prototype of the Berkeley cyclotron was demonstrated successfully. By accelerating atoms through a vacuum and using electromagnets to induce collisions at speeds of up to 25,000 miles per second, the device proved to be a breakthrough in particle acceleration. The researchers continued their experiments throughout the summer, which led to further improvements in the device. Lawrence was kept abreast of all the breakthroughs, with one particularly memorable telegram sent to him in August of that year: 'Dr Livingston has asked me to advise you that he has obtained 1,100,000 volt protons. He also suggested that I add "Whoopee"!'[23]

As scientists developed more powerful particle accelerators, the devices also grew in size. Accommodating the larger cyclotrons required more space, which led to Ernest Lawrence taking over an unused space adjacent to the physics department at

Berkeley. The building was renamed the Radiation Laboratory, or Rad Lab, and became the centre of research for particle physics.

Back in Europe, at the Royal Institute of Physics at Sapienza University of Rome, Enrico Fermi led a team that also aimed to create artificial radioactive isotopes. They opted to use the recently discovered neutron, the reasoning being that since they had no charge, they had the potential to be more effective than alpha particles.

Fermi's team made a crucial discovery. Rather than trying to speed up the neutrons, they slowed them down, i.e. moderated by being passed through a hydrogen-rich material such as paraffin or water. This technique made the neutrons more easily captured, increasing the likelihood of them lodging in the nucleus and causing it to split.[24]

They then systematically tested most of the known elements by bombarding each one with these reduced-speed neutrons. Beginning with the lightest element, hydrogen, they worked their way through the periodic table. For some time, nothing seemed to happen. However, when they reached fluorine, it displayed the desired artificial radioactivity; aluminium was next and that too responded in the same way.[25]

As the research continued, it was regularly reported in the small Italian journal *La Ricerca Scientifica*, which was selected for its quick publication turnaround time. By March 1934 they had published another report which listed twenty more elements that had shown the same effects.

A few months after their already groundbreaking discoveries, Fermi and his team submitted their third report, in which they had reached the heaviest element on the periodic table – uranium. Through experimentation, they found that bombarding it with slow neutrons resulted in the emergence of something new, with their lab's sensing devices detecting that something even heavier than uranium was being produced.[26]

Researchers quickly determined that the neutron-induced substances were not new isotopes of the previously known radioactive elements. As the year ended, they ruled out even more potential suspects, ultimately leading to the conclusion that all elements below Mercury – atomic number 80 – should be eliminated from their search.[27]

Fermi was convinced that a new element had been discovered, possibly a previously hypothesised heavier 'transuranic' element, one that would have a higher atomic weight than uranium.[28] His conviction was so strong that he proceeded to designate a specific spot for the element on the periodic table, positioning it beneath rhenium, a chemical element with the atomic number 75. In addition, he gave it a name that reflected its position, referring to it as eka-rhenium.[29]

These assertions created a stir of excitement and anticipation among the scientific community, and many researchers were quick to accept and support the conclusions. It was widely agreed that the Italian team had likely discovered trans-uranium elements, and multiple research groups, including those of the Joliot-Curies, Lise Meitner and Otto Hahn in Berlin, devoted their time and effort to further study.[30] It seemed that the answer had been found. After all, as Hahn later recalled: 'Everyone else who had worked on the problem had arrived at the same conclusion; there seemed to be no other.'[31]

However, Ida Noddack, a German chemist, was not entirely convinced by the growing scientific consensus surrounding the transuranic discoveries. Noddack was a successful commercial chemist who had transitioned into academic work, though due to marriage laws in Germany at the time she held no formal status and was an unpaid 'work unit' with her husband Walter. She was also a co-(re)discoverer of rhenium, which was one of the last naturally occurring elements to be identified. For this achievement, Noddack had been nominated for the Nobel Prize in Chemistry several times, though she and her colleagues were never awarded the prestigious honour.

Noddack publicly voiced her disagreement with Fermi's hypothesis while giving a presentation at the International Congress held in Russia during September 1934 in honour of the hundredth anniversary of Dmitri Mendeleev's birth. She later published a paper entitled *Uber Das Element 93*, in which she argued that Fermi's team had not conducted enough experiments to support their conclusions.[32]

Noddack also contended that Fermi had misinterpreted the data, and suggested that the answer to the puzzle was that when heavy nuclei are bombarded by neutrons, they break down into numerous large fragments that are isotopes of known elements but not neighbours of the bombarded elements.[33] The answer, therefore, could be found not within the heavy elements, but the lighter ones – the ones Fermi hadn't tested.

Noddack's explanation of the phenomenon was radical. Experience had so far shown that any changes that occurred were minor, and the product nucleus remained similar to the original.[34] Her proposals were not widely accepted and were met with scepticism.

At the time, both the explanation for Fermi's findings and the dissenting views were purely theoretical. Ultimately, there was no way to determine whether it was physically feasible for uranium atoms to split into significant pieces.[35] And until concrete evidence could be obtained to the contrary, the prevailing belief among scientists was that Fermi had indeed discovered transuranic elements, numbered 93 to 97 and perhaps even beyond.

Following in the Italian's footsteps, Lise Meitner, Otto Hahn and Fritz Strassmann at the Kaiser Wilhelm Institute (KWI) in Berlin conducted a series of experiments in 1936, irradiating uranium with both slow and fast neutrons, and varying the duration of exposure. Their findings were consistent with the properties of the anticipated transuranic elements, and as such they remained optimistic about their ongoing research efforts and their place in proving Fermi's discovery.[36]

Their research was to be curtailed by the annexation of Austria by Germany in March 1938, which had a profound impact on the scientific community. This *Anschluss* emboldened Nazis and their sympathisers, both in Germany and beyond. Scientists who either did not support or did not express their loyalty in a suitably enthusiastic way were treated with suspicion. And those who were considered inferior in the hierarchy of races that had come to be one of the defining characteristics of Nazi ideology found their positions and indeed their lives in jeopardy. After the annexation of her country, the Jewish-born Austrian Lise Meitner found herself stateless, passport-less and at the mercy of her colleagues, some of whom immediately denounced her. Chemist Kurt Hess, who later joined the Nazi Party, even raised concerns about the effect of her continued employment at the KWI: 'The Jewess endangers the institute.'[37]

Colleagues in other countries wrote to Meitner, offering fabricated reasons to visit them and escape the growing danger. But without the required permission from the German authorities, she was unable to leave the country officially. She still managed to escape, though, arriving in the Netherlands in July 1938.

She later wrote about the experience:

So as not to arouse suspicion during my last day in Germany, I stayed at the Institute until 8pm correcting the work of one of my younger colleagues. Then I had about ninety minutes to pack some essential items into two small cases and leave Germany for ever with 10 marks in my purse. So I came, in mid-July to Holland without any promise of an appointment; I was almost sixty years of age.[38]

Although no longer in Germany, Meitner continued to collaborate with the exciting research being carried out at her old institution, regularly receiving updates from colleagues, as well as providing advice and guidance. As the year ended, she learned of the results of another variation of the neutron bombardment of uranium. They had again found the expected active products, but this time one had been identified as an element that was chemically similar to radium, but which was almost half the mass of a uranium atom – barium.[39]

Now exiled in Sweden, Meitner had a breakthrough in the understanding of what was happening during these experiments while discussing it with her visiting nephew, Otto Frisch, during the Christmas break. Through mathematical calculations, she realised that the reaction was not caused by the chipping or cracking of the nucleus, as had previously been thought, but was actually the splitting of the nucleus into two approximately equally heavy fragments. Meitner further explained that the loss of mass caused by the split would be converted into energy, which could be measured, and allowed the two nuclei to exist as separate entities.

Meitner and Frisch developed their explanation of the splitting of the nucleus and submitted a report to the editor of Nature.[40] The process had been given a new name, 'fission', a term borrowed from biology which described the splitting of a cell into two. The report was published on 11 February 1939, but unfortunately they were gazumped by Hahn and Strassmann, who independently announced the discovery, using for the first time the term Uranspaltung (uranium fission).

For the Nobel Prize in Chemistry that he eventually received for this work, Hahn failed to mention Meitner's important theoretical contribution, and originally she was not given the credit she deserved within the scientific community.[41] And neither was Ida Noddack, who also received another dressing down after she pointed out, in an article in the journal *Naturwissenschaften*, that she had given her theory to Hahn personally five years earlier and had not received any citations or acknowledgements. Hahn later told her that he hadn't wanted to publicise her assertion because: 'he did not want to make me look ridiculous, as my assumption of the bursting of the uranium nucleus into larger fragments was really absurd.'[42]

This new discovery was quickly broadcast through scientific journals, fuelling further exploration and inquiry. The publication of over a hundred scientific papers, as other researchers replicated and built on the findings, on uranium fission in 1939 alone highlights the widespread sharing of information, which in turn led to significant leaps in knowledge and understanding.

The landscape of international scientific collaboration underwent a significant transformation due to the rise of Nazi Germany. It was not just Meitner who was forced to leave, but other notable nuclear scientists as well. Many had already left Germany in the early years of the decade. For instance, Eugene Wigner relocated to the US in 1930, Einstein arrived in Princeton in 1932, and Edward Teller went to the George Washington University a few years later.

Leo Szilard, a Hungarian physicist, had also migrated to the United States in 1939, where he joined Columbia University and became a part of an impressive research team comprised of many of these European refugees.

Szilard was someone used to coming up with practical solutions to problems that may have commercial applications, especially when they could supplement his income. While living in Berlin years earlier he had come up with many inventions, including the Bride-O-Mat coin-operated mail drop for those who received love letters they wanted to keep a secret, and a new way of keeping stockings from falling down – utilising iron threads woven into their tops and magnets stitched into a jacket pocket.[43] He also worked with Albert Einstein to improve the safety of refrigerators, which were known to leak the toxic gases that were used as coolants. While these never became commercially available, and were soon superseded by other designs, the liquid-metal pumping systems that the two scientists had developed were later utilised by the Manhattan Project.

He also envisioned the potential energy that could be unlocked by spontaneous atomic transmutations.[44] In a much-repeated story, one day in September 1933, Szilard reasoned that it was entirely feasible that at least one of the radioactive elements could absorb a neutron when bombarded, split, and then release two neutrons. These two neutrons could, in turn, trigger a similar reaction in nearby atoms, creating a self-sustaining chain reaction.

As a theory it was intriguing, potentially destructive and absolutely unprovable. But it was not without precedent – in the story *The World Set Free* by H.G. Wells, a scientist converts the substance 'Carolinum' into three atomic bombs which would continue to explode indefinitely.[45] Szilard spent the next few months dreaming about the implications:

> I didn't do anything: I just thought about these things. I remember that I went into my bath – I didn't have a private bath, but there was a bath in the corridor in the Strand Palace Hotel – around nine o'clock in the morning. There is no place as good to think as a bathtub. I would just soak there and think, and around twelve o'clock the maid would knock and say, 'Are you all right, sir?' Then I usually got out and made a few notes, dictated a few memoranda. I played around this way, doing nothing, until summer came round.[46]

The following year, Leo Szilard took out three patents in the UK based on his ideas. One of these, titled 'Improvements in or relating to the transmutation of chemical elements', which in essence described the creation of a neutron-based chain reaction. However, the patent was still purely theoretical, with many missing pieces that would be required to make it work, such as identifying which element would behave in the hypothesised way. When uranium's fissionability was announced in 1938, Szilard, who had already done extensive thinking (in the bath and otherwise, I would imagine) on the matter, was able to quickly grasp the potential of the neutron-induced chain reaction in the element. And at Columbia University, he was well-positioned to continue his work.

Columbia also had Enrico Fermi, another refugee from the rising tide of facism in Europe. The Racial Laws, which sought to restrict the civil rights of Italian Jews, were responsible for the Fermi family being driven out of Italy. Laura Fermi, as a Jewish woman, faced a barrage of severe restrictions and the loss of rights, which also extended to their children. The Fermi family had used his 1938 Nobel Prize in Physics ceremony as an opportunity to escape Italy and start a new life in New York.[47]

For many researchers it wasn't just the possibility of creating a uranium chain reaction that played on their minds, but what would happen to the energy that such an action could release, especially if it was used to create some sort of explosion. Whether or not this was a realistic outcome of a chain reaction was a matter of debate. And while Fermi had, in 1939, shaped his hands into a ball to estimate the amount of uranium that could make Manhattan disappear, the main hope was that such a uranium bomb would remain a theoretical concept, at least for the foreseeable future.[48]

But with the start of the Second World War, the 'what if' factor was a daunting prospect. What if Germany, now isolated from the rest of the scientific community, used its still remarkable expertise to become the only country to develop a uranium bomb?

In the United Kingdom, there were two groups entrusted with assessing the feasibility of such a weapon. One team, led by Professor George Paget Thomson at Imperial College London, had by early 1940 concluded that the idea wasn't worth pursuing further at the present time. In this they were in agreement with Sir Henry Tizard, chair of the UK's Committee on the Scientific Survey of Air Defence, who remarked in mid-1940 that he was 'prepared to take a bet that there would be certainly no military application in this war'.[49]

The other team, headed by Mark Oliphant at the University of Birmingham, working in collaboration with scientists including German/Jewish refugees Otto Frisch and Rudolf Peierls, were much more optimistic. And they set about turning theoretical concepts into reality.

The first hurdle that needed to be overcome was related to the differences in fissionability between the two isotopes of uranium, U-235 and U-238. These isotopes had been successfully separated in September 1939 by Niels Bohr and John Archibald Wheeler, using a mass spectrometer. They were then bombarded with slow neutrons. To the researchers' disappointment, U-238 showed no signs of fission, while U-235 did. This finding was discouraging because U-235 was actually the rarer of the two isotopes. And when I say rare, I'm not exaggerating. In fact, if you were to take a chunk of naturally occurring uranium, over 99 per cent of it would be U-238.[50] It was apparent that uranium enrichment, which involves increasing the proportion of the fissile isotope U-235, would be key to a successful weapons project.

Peierls was tasked with calculating how much U-235 would be required to create a self-sustaining chain reaction so powerful that it would lead to an explosion. While previous estimates had thought that this would be a huge amount, Peierls determined that between one and ten kilograms would be enough to do the trick.

The work of the two physicists was circulated in a two-part document, known (surprisingly enough) as the Frisch–Peierls memorandum, in March 1940. One report, 'On the construction of a super bomb based on a nuclear chain reaction in uranium', detailed how they believed it could be built. The second raised the implications of using such a device with the warning that: 'owing to the spreading of radioactive substances with the wind, the bomb could probably not be used without killing large numbers of civilians, and this may make it unsuitable as a weapon for use by this country.'[51] It was also made clear the importance of keeping all of this top secret.[52]

The memorandum had an almost immediate effect and directly led to the rapid formation of the MAUD Committee, a scientific working group, in April 1940. The group reviewed the evidence and conducted new research into uranium enrichment and atomic weapons design. As they weren't British-born, neither Frisch nor Peierls were allowed to be part of the actual committee.

MAUD's findings, published in July 1941, were actually two reports: 'Use of uranium for a bomb' and 'Use of uranium as a source of power'. Combined, their final conclusions were mind blowing – it was confirmed that making a super bomb using uranium was a real possibility. They also put a timeline and monetary amount on the project, suggesting that it would cost £5 million to construct a plant with the capability to build three bombs per month, with the first being ready by the end of 1943. MAUD also reported on investigations into the potential for uranium to fuel a reactor that could not only be a source of energy, but would also create the new material that had been confirmed to have fissile properties – plutonium.[53]

Because while Fermi hadn't found the elusive transuranic elements, the hunt had continued. By 1940 scientists had theorised – separately, as war time constraints were in effect – that neutrons captured by U-238 would produce an element with an atomic weight of 94, which would be fissionable. While this had been largely theoretical, a team at the University of California at Berkeley actually bombarded uranium with neutrons in the cyclotron and created minute amounts of the radioactive element 93 – neptunium – which in turn decayed into element 94.[54]

And then, in 1941, enough of the pure element had been isolated to confirm that it was not only fissionable but was even more so than uranium.[55]

The conclusions of MAUD were enough to persuade the British government of the importance of research and development into an atomic bomb – a project launched as the Directorate of Tube Alloys – a cover name chosen to be deliberately vague and misleading. While the technology was promising, not everyone was convinced by the utility of such a scheme. The Prime Minister, Winston Churchill, wrote: 'although personally I am quite content with the existing explosives, I feel we must not stand in the way of improvement.'[56]

However, there were a few logistical obstacles to overcome, such as the constant air bombardment of early 1940s Britain, the possibility of a German invasion and also the shortage of workers that was the result of being a country at war. The committee recommended that the project be relocated to either Canada or the United States.[57]

The US had been lagging significantly behind Europe both in atomic bomb research and in recognising the implications of what would happen if other countries were able to utilise the technology for military means. It took a group of refugee scientists in the US to raise the alarm. This group included prominent figures like Leo Szilard. In a letter dated 2 August 1939, written by Szilard but signed by Einstein, this awareness-raising campaign was initiated.[58] The letter

also came with what became known as the Szilard memorandum, which highlighted the importance of securing uranium
supplies:[59]

> The United States has only very poor ores of uranium in mod-
> erate quantities. There is some good ore in Canada and the
> former Czechoslovakia, while the most important source of
> uranium is Belgian Congo. And ... giving particular attention
> to the problem of securing a supply of uranium ore for the
> United States. I understand that Germany has actually stopped
> the sale of uranium from the Czechoslovakia mines, which she
> has taken over.[60]

This was a reference to the town of Joachimsthal, which had been
renamed Jáchymov and incorporated into Czechoslovakia after
the First World War. Now annexed and part of Nazi Germany,
the town became a vital part of the war effort with military hospitals established in the Radium Palace Hotel and other public
buildings. The uranium mines and chemical factory also came
under German control, and during the war were operated by
captured Allied pilots.[61] The idea that the Germans had access
to such a large source of fissionable material was concerning to
say the least.

The report was delivered to Roosevelt on 11 October 1939.
While it had taken some time to get past his advisors, there
was soon action.[62] Roosevelt established an advisory committee to, as he put it in a letter to Einstein, 'thoroughly
investigate the possibilities of your suggestion regarding
the element of uranium'.[63] Lyman James Briggs, director
of the National Bureau of Standards, chaired the committee,
which became known by his name. The Briggs Committee
also comprised military and naval representatives, as well as
scientists like Edward Teller, Eugene Wigner and Szilard.[64]

The committee had differing opinions on the feasibility of creating a chain reaction in uranium, which was still a theoretical concept at the time. And, as a whole, they were also more interested in its potential as an energy source – for propelling a submarine, for example – rather than pursuing the development of a bomb.[65]

It was the final MAUD report, which was properly considered in December 1941, that spurred the US into action. A new controlling body, OSRD Section on Uranium or the S-1 Committee, were tasked with the remit to work out whether it was possible for the United States to make a bomb and to determine at what cost.[66]

The urgency of the situation increased dramatically when the Japanese launched a surprise attack on Pearl Harbour, officially thrusting the United States into war. It took the S-1 Committee only six months to release their report, which stated that building an atomic bomb would require around $100 million, construction of production plants for uranium preparation, and a separate location for bomb development. The report predicted that if all these conditions were met, it would be possible to build an atomic bomb in time to be used during the current conflict; indeed, to end it.[67]

In June 1942, the various streams of atomic efforts were merged into one, codenamed the Development of Substitute Materials but later renamed the Manhattan Engineer District, the location of its first headquarters, or the Manhattan Project for short.[68]

The project was led by the soon-to-be-promoted Colonel Leslie R. Groves, with the scientific component managed by Robert Oppenheimer, who began assembling a team to execute the project's objectives. And the operation had only one objective, really – to build an atomic bomb.

The responsibility for designing a process to create a sustainable chain reaction fell to Professor Arthur Holly Compton. In 1942, research was relocated to the University of Chicago's Metallurgical Laboratory, as were many scientists including Fermi.

The building of an experimental reactor, retrospectively nicknamed CP-1, began on 16 November 1942. Chicago Pile #1 was located right in the middle of the university campus, underneath the west stand of Stagg Field, their football stadium.[69] Using knowledge already gleaned from previous experiments, the reactor was constructed – literally piled up – by students, local labourers and members of the Met Lab. They used almost 400 tons of moderating black graphite bricks, some of which had holes drilled in them to fit fuel 'slugs' – natural uranium in the form of uranium metal or uranium oxide. Other bricks were drilled to fit fourteen-foot-long wooden-handled rods. When inserted, these 'control rods' would absorb a significant number of neutrons, thereby preventing the initiation of a chain reaction. This was thanks to the properties of the cadmium that they were coated with. This safety mechanism was locked in position at the end of the working day and only two scientists, Herbert Anderson and Walter Zinn, had the keys.[70]

None of the previously assembled piles had been large enough to sustain a chain reaction, so there was a question about how big CP-1 needed to be to reach this minimum mass. So, they kept building it, layer upon layer. Block by block. All the while carefully monitoring the pile using a variety of instruments, including a boron trifluoride detector, which was inserted after the fifteenth layer.[71] This vital piece of kit was constructed by the only woman physicist on the Chicago team, 23-year-old PhD student Leona Woods. Her counter measured the activity of the neutrons in the pile – if they were growing exponentially then it was reaching full power, and the team would have to implement one of their control measures to moderate it.

After just over two weeks, when the 57th layer had been reached, calculations indicated that the crucial moment was approaching. On 2 December 1942 it was time to make the pile active. While closely monitoring the instruments, the control rods were slowly removed. They were not needed, but there were plenty of safeguards in place. One scientist, Norman Hilberry, stood on the viewing balcony with an axe, ready to cut a rope that would release an emergency control rod into the pile. Additionally, a bucket of cadmium salt solution was kept on hand that could be poured in as a last resort.[72] Both of these backups would absorb neutrons. Incidentally, the story that Fermi referred to Hillberry as the 'Safety Control Rod Axe Man', which when shortened became SCRAM, the word used for the emergency shutdown of a reactor even today, seems to be just a myth. Unfortunately, there seems to be no definitive etymology for the term, just a lot of semi-plausible theories.

It took several hours to inch the reactor towards full power, and Fermi insisted that they all have a good lunch before the final push. Eventually, it happened – the world's first controlled, self-sustaining fission chain reaction began. It lasted for a mere 28 minutes, during which time only a small amount of energy was produced, less than half a watt. Fermi later wrote: 'The event was not spectacular, no fuses burned, no lights flashed. But to us it meant that release of atomic energy on a large scale would be only a matter of time.'[73]

Leona Woods later recollected the muted celebrations of the 40-odd scientists present and the bottle of Bertolli Chianti that was produced by Eugene Wigner, decanted into paper cups and drunk: 'No toast, nothing, and everyone had a few memorable sips.' The bottle's straw wrapper was autographed by those present and kept to memorialise the event. It's still in Chicago today, although most of the signatures are now terribly faded.

Despite this being a major scientific advancement, there were no news headlines to report the results – instead, as befitting such a secret experiment, Professor Compton sent a simple coded message to James B. Conant, the head of the wartime scientific efforts in the US: 'You'll be interested to know that the Italian navigator just landed in the new world.'[74]

The experiments with CP-1 continued, with the team eventually reaching a maximum of 200 watts by 12 December.[75] The researchers opted not to go any higher with the power output because the pile had deliberately been designed with no radiation-protection measures in place and they wanted to ensure the safety of the scientists present.

The crucial aspect was that they had demonstrated that the system worked. It wasn't important that they had only produced a small amount of energy, as that wasn't the intended purpose of CP-1. Instead, it had been built to 'breed' plutonium. However, although CP-1 was a success as a scientific experiment, it was constructed on such a small scale that recovering significant amounts of plutonium from it was impractical. The production rate was clearly not fast enough to meet the requirements of the atomic bomb project. With the small amount of energy that CP-1 produced it was estimated it would take 70,000 years to produce enough plutonium to create one bomb.[76]

Plutonium was important to the bomb project because it was, along with uranium and thorium, fissionable.[77] The Manhattan Project had decided to focus on exploring two pathways – enriching uranium and breeding plutonium, using various teams and locations across the US.

The work undertaken was shrouded in the utmost secrecy, with a strict emphasis on compartmentalisation and knowledge being on a 'need-to-know' basis. Not only were the sites designated with codenames, but also the personnel involved, and the various components and materials used in the project.

The first issue that needed to be resolved was to secure supplies of uranium. By the early 1940s, the radium industry was in turmoil. The US market had been severely affected by cheap imports in the 1920s, and, in turn, the mines at Shinkolobwe and the Canadian mines at Port Radium had also pulled their shutters down as the market for radium effectively dried up due to concerns around its safety and new advancements in medical technologies.[78] With radium mines, mills and processing sites largely closed, access to uranium became more difficult.

Britain and the United States formed a procurement partnership and established the Combined Development Trust (CDT). The Declaration of Trust was signed on 13 June 1944 'to secure control of uranium and thorium' within the participants' own territories and also to sign contracts with third countries like South Africa. These separate agreements were kept secret, lest the suppliers discovered the price disparity between them.[79] The uranium was to be bought jointly by the CDT and then allocated to the United States and Britain by a mechanism known as the Combined Policy Committee. During the war it was agreed that all uranium would be allocated to the US 'for the production of weapons for use against the common enemy'.[80]

The following year the CDT entered into a contract known as the Belgian-British-American Tripartite Agreement with the Belgian government, which was still in exile at that time. This contract granted exclusive rights to the uranium mined in the Belgian Congo. As part of this understanding, the African Metals Corporation, an affiliate of Union Minière, received substantial funds amounting to millions of dollars to expand and modernise their operations with the aim of maximising uranium extraction.

With supplies of uranium secured, the main challenge faced by the Manhattan Project was to determine an effective process for uranium enrichment. The project undertook an exhaustive investigation of various techniques to obtain substantial amounts of pure U-235 metal, enough to be measured in kilograms rather than trace amounts.

A uranium enrichment plant, codenamed Site X, was built in Oak Ridge, Tennessee, where an entire town was constructed to accommodate the tens of thousands of workers involved in the project. There were three colossal facilities, namely Y-12, K-25 and S-50 to facilitate the two uranium enrichment projects being explored at the site. One of these, labelled as Path 1A, centred on an electromagnetic separation plant located at site Y-12.[81] Electromagnetic separation was the most developed of the various methods at this point in time, and it was hoped it would be able to produce the greatly enriched U-235 almost immediately, albeit in small amounts. At Y-12 they used calutrons, devised by Ernest Lawrence and based on his earlier cyclotron, to separate the different isotopes using a powerful magnetic field. The distinct paths taken by the lighter and heavier isotopes made it possible to collect and isolate them separately.

One of the key elements of this technique was to maintain a constant magnetic field, which required skilled operators in individual cubicles, sitting on stools at six-foot intervals, typically working six days a week for about ten hours a day.[82] It was their job to watch a voltage dial and adjust the knobs to keep the needle pointing straight up, correcting the slight voltage variations that occurred.

These operators are now known as the 'Calutron Girls' and were mostly young women who were just out of high school. This level of education was deemed important, as their training focused solely on the movements essential to their specific job, without delving into the scientific principles behind it. This was to ensure the maintenance of secrecy, but also because it was felt that more educated people might overestimate their technical knowledge and try to fix the problems rather than just turning the dial as requested.[83]

While they were able to enrich the uranium as desired, it was a wasteful, expensive way of doing it, plagued with problems.

Also at Oak Ridge was the K-25 plant, which was focused on using the gaseous diffusion process. There had been lots of experiments with this process, but no one was sure how it would work on a scale as large as that required by the Manhattan Project.

The gaseous diffusion process involves turning the solid metal into uranium hexafluoride, a gas that has one uranium atom and six fluorine atoms, which is pumped through a barrier material that has millions of teeny-tiny holes in it. Because the U-235 atoms are lighter, they can pass through the holes in the barrier more easily than the heavier U-238 atoms, enriching the gas with more U-235. And while this sounds relatively simple, at K-25 there were 3,000 repeats of this process for the uranium hexafluoride gas to pass though. Each step meant they were getting a higher and higher concentration of U-235.

There was also a catch. Uranium hexafluoride is super-reactive and can dissolve most materials it comes into contact with, which makes finding the right gases for the process quite a challenge.[84] And the huge K-25 plant, which at the time was the largest building in the world, had to be made from nickel, the one metal that doesn't react to uranium hexafluoride.[85]

Ultimately, K-25 proved that while gaseous diffusion was a useful tool in the laboratory, the technology at the time was insufficient to work on such a large scale as required. It was a more efficient process than electromagnetic separation, but it also took a very long time to produce small amounts of not-very-high-grade uranium.[86] By late summer 1943 it was clear that K-25 could not produce the weapons-grade uranium that was needed.

Ultimately, by the end of 1944, the uranium enrichment process necessitated utilising many of the known techniques, with each step gradually increasing the concentration of U-235. First, in a building codenamed S-50, the raw uranium was enriched slightly through thermal diffusion, reaching less than 2 per cent. Then the material was sent to K-25, where the enrichment level was raised to around 20 per cent. Finally, the enriched material was fed into Y-12 for the last cycle of enrichment, reaching 80 per cent purity.[87]

Despite the imperfections in the uranium enrichment processes, the goal had been achieved. The enriched uranium was then allocated for different purposes, with some going to a highly classified location in New Mexico and the rest being used to breed plutonium.[88]

The story of the plutonium pathway was very different, but it also started at Oak Ridge. There a pilot reactor, X-10, was built – an immense graphite block measuring 24 feet on each side. Inside, protected by high-density concrete, were 1,248 horizontal, diamond-shaped, air-cooled channels, housing rows of cylindrical natural uranium slugs. When needed, fresh slugs were inserted into the front channels, while the irradiated ones dropped from the back wall through a chute and into an underwater bucket. After several weeks of underwater storage to allow for the decay of radioactivity, the slugs were transported to the chemical separation building, which was also on site. While the process of separating plutonium from uranium was much easier than enrichment, it was still complex and required a variety of steps, including crystallisation, distillation and whirling in centrifuges.[89] And even after that there was still the very important step of further concentration and the removal of impurities.

X-10 began operating on 4 November 1943 and shortly after began to produce small amounts of plutonium.[90]

And while the pathway had been started at Oak Ridge, it was at Hanford – site W – where much of the plutonium work would be done. The leaders of the Manhattan Project knew right from the beginning they would need a huge space, and General Groves had very particular requirements for the facility they were going to build. It needed to be well away from urban areas, with plenty of access to cooling water and ample electrical power.[91]

Like other sites of the Manhattan Project, the location that was chosen for Hanford was already inhabited. So the government took control of a staggering tens of thousands of acres, which resulted in the disappearance of entire towns and a loss of access to traditional homes. The Wanapum Native American nation, for example, had no choice but to relocate. The people who lived there were given a mere 90 days to pack up and abandon their homes and land, forced to leave their lives behind, and mostly with little compensation.[92]

On 6 April 1943, ground was broken at the Hanford facility, marking the beginning of its construction.[93] The Hanford reactors were designed to build on the lessons learned from CP-1 and X-10, but were bigger and more powerful than their predecessors. Unlike CP-1, Hanford used water cooling to effectively manage the substantial heat generated during operation. With approximately 30,000 gallons of water pumped through the reactor every minute, this crucial system played a vital role in ensuring safe and efficient operation. The proximity to water was of utmost importance, and as such, three reactors, labelled as B, D and F, were constructed at six-mile intervals along the Columbia River.[94]

Reactor B, the first on the site went online in September 1944 and reached full power in February 1945, with the first shipment of pure plutonium nitrate sent to New Mexico, to be converted into metal, the same month. The waste products, giant vats of radioactive chemicals, were left on site. By the summer of that year, it was possible to report that the process was operating 'with better efficiency than had been anticipated'.[95] Finally, there was enough uranium and plutonium to complete the Manhattan Project's final objective.

Under the direction of Robert Oppenheimer, the mission at Los Alamos, New Mexico, was to turn the plutonium and enriched uranium into bombs.

The most crucial aspect of this part of the project was actually working out the ways of creating an explosion. So much of the previous experiments had been about controlling the fission process; now they had to make an uncontrolled reaction.

Otto Frisch, now also in the US, was entrusted with the task of determining the size of uranium required for this process. By April 1945, Frisch had completed his experiments and reported the results.[96] This was the final calculation that was needed to build the uranium-based bomb.

The device was given the codename 'Tall Man', though it was later changed to 'Little Boy'. It was a gun-type design, where small, subcritical pieces of pure uranium metal were positioned at each end of a long tubular gun. Upon firing the gun, one piece of uranium was propelled towards the other at an incredibly high speed, like a bullet. When the two pieces collided, they would merge into a single mass, become supercritical and cause the desired explosion.

There was confidence in its efficacy, and due to the immense challenge of producing enriched uranium, it was decided it would be a waste to conduct a test. Instead, the scientists made only one unit, designated as L-11, which would be used in the eventual mission.[97]

The bomb that utilised plutonium posed a different challenge altogether. The gun design was not effective, and scientists had to come up with a new approach. Their solution was to design a method based on the implosion of a hollow sphere of plutonium. Due to the unique nature of the implosion design and the use of the incredibly dangerous plutonium, it was deemed essential to conduct a test of the bomb codenamed 'Fat Man'.

The test, codenamed Trinity, also came with its own challenge. The 'gadget', the first of three nuclear bombs to be exploded by the Manhattan Project, had to be transported from Los Alamos 200 miles away to the Jornada del Muerto basin testing site in New Mexico. Once on site it was hoisted to the top of a hundred-foot tower designated as ground zero.

The test took place on 16 July 1945 at 5.29am Mountain War Time, the explosion leaving a crater with a diameter of around a thousand feet. It is clear that no one really knew what to expect. One of the most poignant recollections came from physicist Norman Ramsey:

> There were a dozen or more of us, in a line up against a bank, with welders' glasses that had been issued to us – we were also supposed to look in the opposite direction when it went off, through the welders' glasses. I must admit I thought that was going too far ... being quite fearful that I might see nothing whatsoever. As it turned out, it was plenty bright.[98]

But if the protection for the assembled scientists was rudimentary, it was still more than that afforded to those who lived in the vicinity. While the exact effects of what would happen when an atomic bomb was set off were not known, it is not true to say that the scientists were completely in the dark to its implications. It was known, for example, that radiation was dangerous – many scientists who had experimented with radium and X-rays in the early days could attest to that. Even more worryingly, calculations conducted only a month before Trinity had confirmed what the Frisch–Peierls memorandum had warned: that an atomic explosion could spread radiological matter across wide areas given the right conditions. With an evacuation order out of the question, due to the need for secrecy, the mitigations put in place involved trying to time the Trinity explosion with wind conditions that would blow the radioactive materials high into the atmosphere and away from the nearby ranches and townspeople.[99]

But while the test was conducted in secrecy, there were many observers. Their reminiscences all make for fascinating reading. From Ramsey's to Oppenheimer's later recollection of how his mind turned to Hindu scripture, to scientist Kenneth Bainbridge's 'Now we are all sons of bitches' – the power of what they had witnessed was palpable.[100] Not all of the witnesses were officially sanctioned by the Manhattan Project, and there were reports in newspapers of a mysterious, unexplained explosion. When pressed, the government, through official journalist William Laurence, intentionally mislead the public with the excuse of an incident at an ammunition depot in the area.

The actual explanation was only a few weeks away. Early in the morning of 6 August 1945, 'Little Boy' was detonated above the city of Hiroshima from the modified B-29 bomber Enola Gay. Three days later, on 9 August at 11.05am, the plutonium-based and slightly larger 'Fat Man' was imploded over Nagasaki, dropped from the B-29 bomber Bock's Car. Both cities were practically reduced to rubble – only five buildings in the entire centre of Hiroshima survived the blast. The intense heat and scorching winds caused by the explosions burned the skin and flesh off anyone within a mile of the detonation area, effectively vaporising them. While the final figure will never be known, according to the United States military approximately 70,000 individuals lost their lives in the Hiroshima bombing, although subsequent independent evaluations have suggested that the actual death toll was 140,000. At Nagasaki the total mortality rate was reported as being 70,000.[101] For those who survived the explosion, the aftermath was also horrific. The radioactive fallout, literally black rain that fell from the sky a few hours after, was a deadly consequence. Those who survived were known as Hibakusha, and they suffered greatly from both the physical and psychological impact of the bomb.

Controversy remains today over the decision not only to use an atomic bomb, but to target those two cities, which had large civilian populations. Indeed, it wasn't a universally supported idea in the first place. Shortly after the Trinity test a group of 70 scientists, led by Szilard, had petitioned President Truman to carefully consider the 'moral responsibilities' of using this weapon against Japan, arguing that the initial stimulus for building them – the threat of Nazi Germany – had been neutralised by the victory for the Allied forces and the end of fighting in Europe. The scientists warned about the immense destructive power of atomic bombs and urged the United States to exercise restraint and take on the responsibility of preventing global devastation.[102]

After travelling to Washington, DC, Szilard personally delivered the petition to Secretary of State James Byrnes, hoping it would reach President Truman. Unfortunately, Byrnes was not sympathetic and failed to pass it on. As a result, the petition was filed away, having no impact on the decision-making process about the use of atomic bombs. The document remained classified until 1961.

With Gallup polls in 1945 indicating that only 17 per cent of Americans saw the bomb as a 'bad thing' and a *Fortune* magazine survey showing that 22 per cent of respondents wished more bombs had been dropped on Japan, how the launch of an atomic bomb manifested itself culturally is interesting.[103] There were bad-taste jokes, as well as speculations that the word would permeate the language: 'What a lot of things will now become atomic, and actions too. There will be the speaker with the atomic delivery, the writer with the atomic expressions ... Titanic, gigantic, dynamic, have become rather good superlatives, but atomic – why, it's simply colossal.'[104]

There was even a plethora of atomic and celebratory cocktails. Those served at the Washington Press Club were made of a mixture of gin and vermouth with a dash of Pernod or absinthe. Partakers of which, apparently, all agreed 'that it could destroy all forms of human life'.[105] At Trader Vic's in Los Angeles the celebratory 'A-bomb' cocktail served was a mixture of rum, blue curaçao and dry ice to make it bubble and smoke. When Mary Lawrence, wife of Ernest, tried one at a Victory over Japan Day party, she proclaimed it: 'Ghastly!'[106]

Burbank Burlesque Theatre, Los Angeles, featured an illustrated advert for 'Atom Bomb Dancers' with headliners Diana Van Dyne and Dixie Sullivan. And in September 1945, *Life* magazine featured a two-page photo spread of 22-year-old starlet Linda Christian. In this profile she was tastelessly proclaimed the 'Anatomic Bomb'. While most of the images are typical starlet cheesecake photos of the period featuring her in a two-piece swimming costume, there is one that shows Christian lying on the side of a pool with a limp arm trailing in the water, her head fallen to one side and unfocused eyes. The caption 'soaks up solar energy' is rather at odds with the death-like pose of the model.[107]

And while the focus was on the effects of atomic bombs, uranium's role in the development was not forgotten. Norman Corwin, a renowned writer of radio dramas, created a special broadcast called *Fourteen August* to commemorate VJ Day and mark the end of the Second World War. The original broadcast aired on 14 August 1945 and was fifteen minutes long. However, due to its significance and popularity, he was asked to provide additional material.

Corwin quickly extended the broadcast, now renamed *God and Uranium*, which aired the following Sunday. Orson Welles was the narrator for both episodes, and the actress Olivia de Havilland joined him for the longer version. The broadcast, with explosion sound effects in the background, left the listener in no doubt as to the role of the atomic bomb in the ending of the war:

> Congratulations for being alive and listening on this night, millions didn't make it. They died before their time and they are gone and gone. For the fascists got them. They are not here, but their acts are here and they are to be saluted from the lips and from the heart before the conversation swings around to reconversion; fire a cannon to their everlasting memory.
>
> God and uranium were on our side. The wrath of the atom fell like a commandment and the very planet quivered with implications.[108]

ATOMIC PREDICTIONS 4

The Manhattan Project had cost around $2 billion and, at its peak, employed 130,000 individuals across 30 sites, most of whom had no idea about the true nature of what they were contributing to.[1] The secrecy surrounding the project also meant that the world was unprepared for the reality of atomic power.

However, some groundwork had been laid thanks to reports on fission experiments and the potential of the atom – focusing on energy generation and other beneficial applications in the 1930s. Investigations into radium had also played their part, especially an intriguing observation made by Marie Skłodowska Curie and Pierre Curie. During experiments they had noticed that their samples of radium chloride not only gave off a faint glow but consistently registered a slightly higher temperature compared to their surroundings, although they were unsure about the source of this energy at that time. This led to speculation about the possibility of using radioactivity as a sustainable source of heat. At the time several journalists pondered the idea of homes being able to eliminate traditional fires and instead rely on a chunk of radium that could provide long-lasting warmth.[2]

These speculative futures extended to predictions made by scientists engaged in research into the properties of the radioactive elements. From Pierre Curie estimating that radium would be able to melt its own weight in ice over the course of a year, to the more specific predictions of Frederick Soddy, who in his 1906 lecture 'The Internal Energy of the Elements' predicted that £1,000 of uranium could be used to produce more energy than was currently generated by 'all the electric supply stations of London put together'.[3] Some scientists were so certain of the inevitability of this technological progress that in 1920, Sir Oliver Lodge was quoted as prophesying that 'the time will come when atomic energy will take the place of coal as a source of power'.[4]

Descriptions and predictions of the benefits of atomic energy increased as atomic power transformed from speculation to reality. Especially in the pages of popular magazines, which envisaged uranium's role as a real alternative to the energy available at the time – coal, electricity, gasoline – but more cost effective and more powerful.[5] For physicist Rudolph M. Langer in an article titled 'Fast New World', a 'new and fascinating way of life' could be in humankind's grasp though uranium. While some of these predictions, like light being generated by uranium and piped under the house through transparent plastic sheets, were incredibly speculative, there were plenty of things to get the reader interested in the potential benefits of uranium.

As the various atomic projects gained momentum during the Second World War, what was later characterised by the editor of *Popular Mechanics* as a 'cloak of silence' had descended over developments in atomic research.[6] This was enforced through censorship measures: in July 1943 the United States Office of Censorship communicated to daily newspapers, weeklies and radio stations that terms like 'atom splitting', 'atomic energy' and 'nuclear fission' should be avoided in articles and broad-casts. Additionally, details about any radioactive elements, their components and materials or equipment being used in atomic research, such as the neutron-moderating heavy water or cyclotrons, were also now prohibited from being discussed or published.[7]

Despite the censorship restrictions, some companies contin-ued to explore the topic of atomic energy in their publications. These types of speculations were so prevalent that in September 1944 a report listed 77 violations that had been escalated to the Office of Censorship (OoC).[8]

One of the breaches that was brought to the OoC's atten-tion was the short story 'Deadline' by Cleve Cartmill, published in *Astounding Science Fiction* in March 1944.[9] The piece of science-fiction writing was pretty run of the mill, but there was one thing about it that had really drawn attention: its accurate description of top-secret science projects.

At Los Alamos, Edward Teller recalled that the story had 'provoked astonishment in the lunch table discussions'. At Oak Ridge, a lieutenant of the Intelligence and Security Division wrote to Lieutenant Colonel John Lansdale of the Military Censorship Department in Washington to raise their concerns about the piece on the grounds that it may 'pro-voke public speculation' into the top-secret work being carried out.[10]

Thanks to US Army Intelligence declassified files we know that an investigation was launched into how a writer like Cartmill had such detailed knowledge of isotope separation and atomic bomb construction and how this was allowed to be published. John W. Campbell, the editor of the magazine, was visited by agents from the Counter Intelligence Corps and took full responsibility for the technical information in the story. It was, he argued, perfectly possible for anyone who understood science to have put two and two together from the publicly available information.

Campbell's established background in science lent credibility to his successful defence against the charges, as did *Astounding Science Fiction*'s track record of publishing stories about atomic energy, including Robert Heinlein's short story 'Blowups Happen', which featured a character named Dr Silard and uranium explosions.[11]

The OoC ceased its operations in August 1945, but that did not signal the end of efforts to manipulate the public's understanding of atomic weaponry. The United States government had effectively kept the secret of the Manhattan Project. During the war only a few scattered pieces of information emerged in public. In the aftermath of the deployment of the atomic bomb it was considered prudent for policymakers to continue this level of secrecy.

One of the ways this was accomplished was through government-sanctioned information releases, such as the official account of the Manhattan Project, which was published shortly after Hiroshima. This report, titled 'Atomic Energy for Military Purposes: The Official Report on the Development of the Atomic Bomb under the Auspices of the United States Government, 1940–1945' and authored by Henry DeWolf Smyth, became known as the Smyth Report.

In addition, journalist William Laurence played a significant role in shaping public perceptions at the time. He was granted unprecedented access to classified information by Leslie Groves, who appointed him as a 'special consultant'. Laurence was allowed to visit the top-secret Oak Ridge, was present at the Trinity test and even witnessed the bombing of Nagasaki from the air.[12]

In exchange for this exclusive access Laurence wrote a number of government-approved propaganda pieces, which were only published after the war had ended. These stories, along with the Smyth Report, significantly influenced how the American public learned about the bomb, shaping their perceptions of the weapon, while at the same time controlling information released to preserve military secrecy. They also sought to debunk the growing rumours, published in papers such as the *London Daily Express*, about people in the aftermath of Hiroshima suffering from an unexplained wasting disease, while also minimising the devastation that had been caused by the bombs on the Japanese cities.[13]

In September 1945, *Time* magazine reported on a tour of the Trinity site by 31 journalists 'to see with their own eyes the first awesome footprint of man's newest genie in the earth of New Mexico'. This excursion, which was arranged by the US Army and led by General Groves and Robert Oppenheimer, was a propaganda visit designed to convince the American public of the safety of the site, and thus by extension scotch any rumours of the lingering radiation at Hiroshima and Nagasaki. Not only were they encouraged to view the site, but also to take away a souvenir of their visit, specifically 'crater glass'.

Crater glass was another name for atomsite, which was also known as Alamogordo glass or trinitite. This is a glassy mineral that was created by the intense heat of the explosion of the Trinity test reacting with the sand. The heat liquefied the sand, and when it cooled and solidified there was a new substance. The majority of trinitite created was a greenish colour, but there were also varieties of red which contained material from the copper wires of the bomb itself and a very rare black form which is believed to have been fused where the tower had stood.

Time told of the journalists who 'enthusiastically pocketed' their souvenirs, which were noted to be still very mildly radioactive but generally safe to handle. It also said that some of the reporters had second thoughts, especially 'when someone recalled the effects of radiation on fertility'.

One of the reporters present that day was Stuart Dixon, the president of Transradio Press. His souvenir was used in a further piece of propaganda: atomic jewellery. In September 1945 designer Marc Koven, of the company Koven Frères in New York, was commissioned to create jewellery using trinitite. This range was to include hair adornments 'designed in a cage-like form with the atomsite as the center symbolizes the nuclear theory' and earrings. Actress Merle Oberon was reportedly the first to wear the pieces, but many of the accompanying images and press releases featured the model Patricia Burrage. The range was also exhibited as part of the United Seamen's Service 'Night in Paradise' exhibitions held in New York and other cities as part of a drive for the National War Fund to raise $113 million. The jewellery was seen as part of the ushering in of the atomic age, along with other post-war products such as cars and refrigerators, also displayed at the exhibitions. It also specifically referenced the bomb: 'The piece may also be described as a jeweller's conception in miniature of an actual atomic explosion, mushrooming skyward atop a pillar of cloud leaving glassy green devastation behind.'

That it was specifically propaganda is also made explicit by some of the covering reports who agreed that the fact that atomsite was safe to wear would refute 'the Japanese claims that it is radioactive long after an atomic explosion'. Thanks to a carefully cultivated message about its role as a decisive factor in ending the war and the myth of how many lives it had actually saved, the early public responses, overall, to the atomic bomb can be characterised as one of celebration coupled with a callous disregard for the lives of the recent enemy.

While there were those who immediately expressed their horror at the atomic developments through misinformation, dismissal and just plain censorship, it wouldn't be until 31 August 1946 that many Americans would start to understand the full extent of the events that had taken place in Japan, through a special edition of the *New Yorker* magazine. Written by John Hersey, the report vividly recounted the story of the destruction through the eyes of six atomic-bomb survivors, including two doctors, a Catholic priest, a Methodist minister and two working women. Spanning 68 pages, the article was simply titled 'Hiroshima'. Its first paragraph captured precisely the mundanity of life before the horrors of the atomic bomb were unleashed:

> At exactly fifteen minutes past eight in the morning, on 6 August 1945, Japanese time, at the moment when the atomic bomb flashed above Hiroshima, Miss Toshiko Sasaki, a clerk in the personnel department of the East Asia Tin Works, had just sat down at her place in the plant office and was turning her head to speak to the girl at the next desk.[14]

The edition sold out and had to be reprinted. Albert Einstein, for example, wanted a thousand copies to distribute to members of the Emergency Committee of Atomic Scientists, a group founded in November 1945 to promote international control of atomic energy, disposal of existing bombs and education into the peaceful uses. When Einstein sent out copies of the *New Yorker*, he included a heartfelt message in his cover note: 'I believe Mr. Hersey has given a true picture of the appalling effect on human beings ... And this picture has implications for the future of mankind which must deeply concern all responsible men and women.'[15]

Despite previously expressing strong support for the use of the atomic bomb, columnists and editors were compelled to praise the article. There was even a rebuke about initial reactions to the bomb for any American 'who has permitted himself to make jokes about atom bombs or who had come to regard them as just one sensational development that can now be accepted as part of civilization'.[16]

∽◦◦◦∾

What do you do with so much power when the war ends?

The implications of this had been considered even before the 'gadget' was ready. A presidential advisory group – the Interim Committee – had been set up in April 1945 to make recommendations and guide the President on policies that would be appropriate in the aftermath of the war.

Various proposals for creating a suitable regulatory agency were introduced in Congress. Interest groups were formed to lobby legislators and there were months of intensive debate between politicians, military planners and scientists. The final bill was a hybrid passed unanimously by the Senate on 1 June 1946 and signed into law by President Truman two months later.[17]

The Atomic Energy Act (AEA) of 1946 included a crucial provision: the establishment of the Atomic Energy Commission (AEC), which was to assume complete control over atomic power, including weapons production, wrapping it again in a shroud of secrecy and breaking all established scientific ties with other countries.

The AEC was led by five commissioners, appointed by the President of the United States.[18] There was also a general manager, again initially appointed by Truman, who would act as a chief executive officer. However, during the transitional period before the AEA came into effect at midnight on 1 January 1947, General Groves retained control of all the resources and decision-making authority. This period, which lasted over a year, was also characterised by a lack of clear policy direction, exacerbated by the end of a long period of war. Groves was left to make crucial decisions for this complex industrial empire that employed some 44,000 people.

As *Life* magazine reported in December of that year: 'Their present job: gradually to take over the vast atomic energy plant built by the Army's Manhattan District. Their higher responsibility: to hold atomic energy in trust until the world can agree on how it should be used without danger to enrich the lives of men.'[19]

One of the initial tasks was to demobilise the Manhattan Project and replace the wartime officers with 'men who were young enough to break into the atomic field, but who were senior enough in rank to have demonstrated their ability to accept heavy responsibilities, and whose age would be an asset in their dealings with our scientific personnel'.[20] Among the many changes was physicist Dr Norris Bradbury replacing Oppenheimer as the director of Los Alamos laboratory. The AEC not only took over the many huge installations that had been part of the Manhattan Project but also the oversight of the contractors that operated them, like the DuPont Company, Union Carbide and Carbon Chemicals Company, and General Electric.[21]

Despite having successfully detonated three atomic bombs, the US was still far from fully understanding the nature and effects of these devastating weapons. A task force was established to develop a comprehensive testing programme, which was approved by President Truman on 10 January 1946.[22] In charge of this joint Army–Navy task force was Vice Admiral William H.P. Blandy, who came to be known as 'the Atomic Admiral'.[23]

Their initial objective was to identify a suitable location for conducting the testing programme. This location needed to meet specific criteria, including being under US control but sufficiently far away from continental America, having a suitable climate and a sparse population.[24]

After careful consideration, Bikini Atoll in the Marshall Islands was chosen as the testing site. This area in the Pacific Ocean fulfilled all the requirements, including having a port area where naval vessels could be anchored.[25] The population of Bikini Atoll was small, with only 167 individuals residing there at the time. The US had been administering authority over the island since its liberation from Japanese occupation in 1945.[26]

Although there are differing accounts on how residents were 'asked' to leave their homeland, it is documented that in 1946, Commodore Ben Wyatt, the American Governor of the Marshall Islands, visited on 10 February. Through a translator, he informed the Bikinian people that they had no choice but to relocate. However, he assured them that they would be taken care of and could return at a later date.[27]

They were offered three relocation sites, and they chose a small uninhabited island called Rongerik, which was just over a hundred miles away from their home. They were transported onboard US Navy landing craft. The new location was completely unsuitable, with temporary housing and limited facilities. They were provided with only a few weeks' worth of food and were left to fend for themselves.[28]

The promises of being taken care of during the testing period and being able to return home after the tests never materialised, leading to long-term displacement and adverse impacts on the community. Not only were the people exiled, but the landscape was altered as extensive preparations were made. This included the need to make space for the ships that would be targeted during the tests. To achieve this, dynamite was used to remove coral heads from Bikini Lagoon, which clearly had a significant ecological impact.

With the Bikinian people relocated and the area cleared, the next phase of Operation Crossroads commenced. The tests were conducted in two parts, known as Able and Baker.

On 1 July 1946, the first atomic test, Able, took place when a 23-kiloton, plutonium, implosion-type nuclear weapon was detonated thousands of feet over the Bikini Lagoon.[29] The bomb itself was nicknamed 'Gilda', and that name was painted on it in bold black letters, along with an image of Rita Hayworth, the star of the 1946 film noir *Gilda*, which was cut out from the June edition of *Esquire* magazine.[30] The *Los Angeles Times* reported this under the heading 'Fission Figure'.[31]

Rita Hayworth's biographer Barbara Leaming later interviewed the actress's then-husband Orson Welles, in which he recalled:

> ... the angriest was when she found out that they'd put her on the atom bomb. Rita almost went insane, she was so angry. She was so shocked by it! Rita was the kind of person that kind of thing would hurt more than anybody. She wanted to go to Washington to hold a press conference, but Harry Cohn, president of Columbia Pictures, wouldn't let her because it would be unpatriotic.[32]

The tests were designed to assess the destructive power and aftermath of atomic weapons in a controlled environment, with the aim of gathering data on the impact of nuclear explosions on ships and living beings, including farm animals that were placed on the vessels. The target for the Able test was the USS *Nevada*, which was painted orange to distinguish itself from the other ships in the central cluster of vessels, including some taken from the defeated enemy.[33] The *Nevada* held significance as the only battleship to get underway during the attack on Pearl Harbor in 1941.

However, much to everyone's disappointment, the bomb missed its primary target, and the radiation readings showed the event was not as devastating as had been predicted. Dr David

Bradley, a radiological safety officer, stated: 'it would appear that at least for the time being we have escaped from the real threat of atomic weapons, namely the lingering poison of radioactivity. The great bulk of highly dangerous fission products was carried aloft into the stratosphere where it can be diluted to the point of insignificance in its slow fall-out.'[34]

The tests of Operation Crossroads were conducted openly and reported on in real time by invited journalists, photographed extensively and even filmed. And with more public awareness there came the beginnings of criticisms of the testing, which largely focused on the lack of understanding of the damage that could be caused. For some, Vice Admiral Blandy wasn't an atomic admiral but an 'atomic playboy' who was hellbent on destroying the world.[35]

Of course, with so much publicity around the tests, various references to them arose in popular culture. In 1946, the General Mills Corporation offered a promotional item that could be obtained by sending in fifteen cents and a KiX cereal box top. Essentially a spinthariscope, the atomic ring had a 'sealed atom chamber' in a 'gleaming aluminium warhead', where one could 'see genuine atoms SPLIT to smithereens!' The toy ring also had a space behind the bomb warhead for secrets, with the claim that it could help users outwit enemies by concealing a hundred-word message in the strategic compartment. According to records, a total of 3 million of these KiX rings were made between 1947 and the early 1950s.[36]

The Andrew Jergens Company of Cincinnati, Ohio, trade-marked the label for 'Atom Bomb' perfume. They began selling their fragrance in its distinctive rocket-shaped bottle soon after. This earthy and potent perfume was available for about 25 cents per bottle at pharmacies and stores like Woolworths.

While the ring and the perfume are now confined to our cultural memory in museums and eBay (and the image pages of this book), something with more contemporary significance can be found in the name of a swimwear design. Both the French couturier Jacques Heim and the Swiss engineer Louis Réard are credited with popularising this garment, which they respectively dubbed the 'atome' and the 'bikini'. While two-piece garments have been attested to from at least the fourth century in Italy, these triangular pieces of fabric connected by strings around the neck and back for the bra, and by strings at the hips for the bottom, was something new. Réard introduced his version only a short time after the Able test, and it was first worn by French model Micheline Bernardini at the Piscine Molitor, a Parisian swimming pool. The press release announced: 'Like the bomb, the bikini is small and devastating.'[37]

This daring swimwear was showcased in America by *Harper's Bazaar* magazine, with a Toni Frissell photograph of a model wearing a green-and-white polka-dot bikini by the sportswear designer Carolyn Schnurer, which featured in the May 1947 issue.[38] Referred to as the 'swoonsuit' by legendary fashion editor Diana Vreeland, the garment was considered scandalous and provocative at the time. Vreeland famously declared that the bikini revealed 'everything about a girl except her mother's maiden name'.[39] However, the design faced resistance in many countries for religious and moral reasons. It was banned in countries like Spain, Portugal and Italy, and was frowned upon by American women, including some models, who refused to wear it, seeing it as indecent. Many public parks and beaches prohibited bikinis and wearing them in private clubs and resorts was looked down upon.

Throughout the 1950s, the bikini remained a taboo novelty, with limited acceptance in mainstream fashion. It was not until the 1960s that it gained more widespread acceptance and became a popular swimwear choice for women around the world.

The second in this series of atomic tests, Baker, was conducted on 25 July 1946. The bomb was suspended beneath a landing craft in the middle of the target fleet and was detonated 90 feet underwater, using electric remote control.[40] The detonation resulted in a massive water column, rising to approximately 6,000 feet, 'which then collapsed back into the lagoon, generating a wave that was nearly the height of the Chrysler Building'.[41] Unlike Able, Baker caused significant damage to the targets, which included 87 battleships and submerged submarines.[42]

The chemist Glenn Seaborg, who later became the chair of the Atomic Energy Commission, called Baker 'the world's first nuclear disaster'.[43] The comedian Bob Hope summed up the US atomic testing programme just as succinctly: 'As soon as the war ended, we located the one spot on earth that hadn't been touched by war and blew it to hell.'[44]

In carrying out a programme of nuclear tests the AEC had two main missions, which were often, inevitably, in conflict with each other. Their first objective was to develop tactical atomic weapons, and for this they needed to conduct tests. However, the second objective was to achieve this goal in the safest (ish) way possible. But to ensure that their primary objective was not jeopardised by the second, the AEC believed that it was necessary to keep the public calm and avoid causing too much concern. This led to a period of emphasis on the peaceful applications of atomic energy while also downplaying the potential risks of embarking on such a widespread weapons programme. This effort took the form of a comprehensive public relations campaign that utilised all available media platforms, including films, public speakers, classroom demonstrations and travelling exhibitions.

One of the most popular of these exhibitions was 'Man and the Atom', which originally took place at the Grand Central

Palace in New York's Central Park, from 23 August to September 1948. It was sponsored by many businesses, including General Electric and Westinghouse, prominent energy contractors, and featured a wide variety of displays showcasing the benefits of nuclear energy, as well as its potential applications in various fields.[45]

The General Electric section distributed copies of the comic *Dagwood Splits the Atom*. Prefaced with a letter from General Groves, this 37-page full-colour publication featured Mandrake the Magician serving as the narrator, guiding viewers through the complex science of nuclear physics. The book employed beloved characters such as Blondie, Popeye, Wimpy and others to explain the science in a fun and engaging way, and it was carefully designed to make anyone feel that they could understand and appreciate the wonders of nuclear science.

In the Westinghouse sponsored section was the Theatre of Atoms, which featured several fascinating exhibits. One of these showed a 'real radiation detector at work', which illustrated a chain reaction using 60 mousetraps that set each other off.[46] In addition, visitors could see various displays, including animals that had been fed radioactive sugar, silver dimes bombarded with neutrons to make cadmium, and even radioactive fertiliser.[47] One standout exhibit featured the effects of radiation on blood cells and encouraged visitors to donate blood in preparation for 'any atomic bomb emergencies'.[48]

'Man and the Atom' was a success, with good visitor numbers and a quantifiable outcome. According to a survey conducted at the time, which involved entrance and exit interviews, visitors to the exhibition reported feeling less fearful and anxious, and more hopeful about the peacetime applications of atomic energy.[49]

The exhibition eventually found a permanent home at the American Museum of Atomic Energy in Oak Ridge, which had been renamed Oak Ridge National Laboratory after the war.[50] The museum had opened its doors in March 1949 and organised its displays thematically, highlighting the various ways in which the atom could be applied for the greater good in areas such as energy, medicine, agriculture and international peace.[51]

After enjoying the interactive exhibits, such as using Geiger counters to locate radioactive reptiles or to test samples of radioactive ore, visitors could visit the gift shop for souvenirs.[52] Here you could buy a sample of uranium ore enclosed in a plastic case or a copy of *Dagwood Splits the Atom* for only ten cents. Or perhaps the ultimate in souvenirs: an irradiated silver dime. This was made in a special isotope cabinet and then encased in plastic. The radioactivity was short lived, only lasting for about four minutes. It is estimated that between the opening of the museum and 1969, when it was stopped, over a million of these were irradiated and given out.[53]

URANIUM FEVER

5

> I do not know if you can do anything, but I am informed that
> if the Bill to control Atomic Energy passes in its present form,
> our Works at Walthamstow will be unable to obtain supplies
> of Uranium Oxide – an essential material for the production
> of our new Primrose Cooking glass.
>
> We have built up a large and growing business in this newly
> invented glass and have on our books substantial orders for
> Home and Export, which it will be impossible to execute with-
> out the above mentioned essential material.

By October 1947, R.W. Johnson was one of the growing band of
manufacturers increasingly frustrated with the scarcity of uranium in Britain. Johnson's company used it to produce Duroven,
chemically toughened glass cooking dishes that were attractive
enough to be 'served straight from the oven' onto your dining
table.[1]

Determined to secure a reliable source of uranium, Johnson
took various steps to address the issue. He contacted his
Member of Parliament and sent letters to the Ministry of
Supply, outlining the urgent need for the element in commercial production. He even made a personal visit to the Ministry's

headquarters, housed in the magnificent Shell Mex House, an iconic Art Deco building located on London's Strand.[2] But despite his efforts, Johnson's attempts were unsuccessful. He was informed that there was no uranium available for commercial use at the time, and it was unlikely to become readily available for several years.[3]

As the war continued and reserves continued to dwindle, as revealed in the declassified files now accessible at the National Archives in Kew, the entities responsible for uranium supply, such as the Board of Trade and the Ministry of Supply, were bombarded with letters from companies concerned by the lack of uranium.[4]

Uranium's use in glassware and ceramics had continued into the 1930s and 40s, with popular brands such as the cosmetic company Dubarry introducing their uranium glass talcum-powder flask – which contained talc fragranced with their signature scents 'The Heart of Rose', 'A Bunch of Violets' or 'Golden Morn'.[5] At around the same time Joseph Richard Lillicrap introduced Lillicrap's Hone, a block which was designed to sharpen and extend the life of safety razor blades and was 'made in Uranium Glass with a specially prepared surface'.

A 1947 survey conducted by the Board of Trade outlined various known applications of uranium, including its use in the pottery industry, the production of fluorescent lighting, the manufacturing of starter switches for electric motors and the processing of colour film, to name a few. It was used in scientific equipment, for demonstration purposes and in Crookes eye protective glasses, the forerunner of modern sunglasses.[6]

In fact, the BOT report continued, it was projected that Britain would require approximately 30 tons of uranium annually to meet its industrial needs, with the pottery industry alone indicating a demand for twenty tons.[7]

The Combined Development Trust had achieved control over a staggering 97 per cent of global uranium ore and 65 per cent of thorium ore,[8] but had broader objectives beyond mere access. It was equally important to ensure that uranium was exclusively allocated to produce atomic bombs and research into energy, which necessitated strict restrictions on civilian usage.[9] In the United States this civilian control was facilitated by the WPB Conservation Order M-285 of 1943, and the AEC issued 'Regulation for Control of Uranium and Thorium' in March 1947, which detailed 'beginning April 1, no person may transfer, deliver, receive title to, or possession of, or export, any of the atomic source material after "removal from its place in nature" without a licence from the commission'.[10]

In Britain the Atomic Energy Act (1946) didn't restrict the use of uranium for commercial purposes as long as it didn't involve 'the production or use of atomic energy'; in practice, however, there was just not enough of the element.

There was also a real confusion around permitted uses, the Americans, who ultimately controlled the supplies, making it clear that uranium was only for 'essential use', which was deemed for analytical or research purposes. Eventually, the AEC confirmed:

> In determining whether or not a contemplated use for uranium will be authorised the primary question is: does the presence of uranium lend useful qualities to the product, or is the uranium required for decorative purposes. If the use of the uranium is not for decorative purposes, but a use which serves a functional purpose in the manufacture of a product of general value, then it is likely to be approved.[11]

For Duroven, along with numerous other potteries and glass-making firms, they relied on the uranium to achieve a desirable colour but could not claim that, other than aesthetically, it was important to their product. However, while not useful or functional under the AEC definition, in a post-war Britain it was the colours that could be achieved by the substance that made them marketable internationally, and many companies were facing real financial difficulty. The President of the Ceramic Printers' Association made this clear in a letter, dated January 1950:

> Members of the Ceramic Printers' Association, are receiving complaints that overseas buyers are claiming that the colours used now (instead of those with an Uranium base), are not satisfactory, and the sales of pottery are being prejudiced in the American markets for this reason.
>
> I know that you realise that British Pottery in the United States is sold only by reason of its superiority, and that unless the quality of the wear and of the decoration is above criticism, there is more than a risk that the market cannot be held.[12]

In the United States, companies also found a lack of uranium to be challenging to the normal running of their businesses. One of these was the Homer Laughlin China Company of West Virginia who, since January 1936 under the watchful eye and design brilliance of art director Frederick Hurten Rhead, had produced a successful line of pottery – known as Fiesta Tableware.

Their mix and match colour concept – this was, after all, during the Great Depression, when people couldn't always afford to buy a whole set at once – was a great success. Twelve million pieces in a variety of colours, including old ivory, blue, green and red, were shipped annually at the company's peak in the 1930s and 40s.[13] It was the red glaze that was the most desirable offering, its intense colour achieved by adding uranium oxide. With compulsory confiscation of uranium stocks by the US government during the war and manufacturing shifting to war-related production, the company began to reduce the number of items in its Fiesta line and the red glaze was temporarily retired.[14]

While the CDT partnership and restrictions had secured the supply chain of uranium, the United States grew increasingly concerned about an excessive reliance on overseas sources, recognising it as a looming threat to national security.[15] During the Second World War this reliance had been at its most acute, with an estimated 85 per cent of uranium used by the Manhattan Project coming from imported sources.[16] And the US enriched uranium that had contributed to the project had essentially wiped out their stockpile.

Access to the main overseas sources, Eldorado and Union Minière, presented their own set of challenges. Eldorado, which mined pitchblende from Port Radium, was situated less than 30 miles from the Arctic Circle, and transporting the uranium required traversing over 1,200 miles of lakes and rivers before reaching the railroad. Similarly, Shinkolobwe posed its own transportation difficulties, with an arduous journey from the mine involving a 1,400-mile trek by river and rail, followed by a perilous 7,000-mile voyage across the Atlantic.[17]

These issues, combined with still slow plutonium production, meant there was real concern about the ability of the US to maintain its dominance over atomic power. By April 1947, a report was presented to President Truman confirming that the atomic arsenal of the United States comprised of fewer than twelve unassembled bombs.[18] This highlighted a 'certain serious weakness in the situation from the standpoint of national defence and security'.[19] The report coincided with Congress's efforts to reshape the military in response to growing Cold War tensions, as reflected in the National Security Act signed on 26 July 1947. Among many changes enacted, the newly created Department of Defense took a more comprehensive approach

to considering the use of atomic weapons for military purposes, including drawing up lists of Soviet targets that could
be attacked.[20]

Shortly afterwards, the Joint Commission on Atomic Energy
(JCAE), the overseer of the AEC, in agreement with the military, emphasised the strategic importance of increasing weapons
production.[21]

And, of course, a crucial part of this was to increase the
amount of uranium available. The Soviet Union's supplies
had been secured through its control of Jáchymov. At the end
of the war the Czechoslovakians had retaken the area and
expelled the German population. A confidential memorandum of understanding between the governments of the Soviet
Union and Czechoslovakia regarding exploitation of uranium
was signed in November 1945, which left the Russians in
control of the mines. The workers of these mines were largely
prisoners, first German prisoners of war and then later crim-
inal and political prisoners. Until the early 1960s, tens of
thousands of these prisoners were put to work in such unbear-
able conditions that the area became known as Jáchymov Hell,
all to supply the Soviets with enough uranium to keep up in
the arms race and develop their nuclear energy capability. F-1,
the Soviet's first nuclear reactor, was eventually powered up
on Christmas Day 1946 using technology largely stolen from
the Manhattan Project by Klaus Fuchs, Julius Rosenberg and
other spies. The quest for uranium increased, at a frenetic
pace, and the industry flourished as the arms race escalated
and increasing amounts of potential fissionable materials were
needed.

In a clever move to avoid over-reliance on overseas shipments, a decision was made to establish a domestic uranium
programme. This was coordinated by the AEC from an office in
Grand Junction, Colorado.[22]

It was to be a uranium boom.

By the close of 1947, the Raw Materials Division was formed under the control of the AEC, with a singular mission: to procure uranium. The programme's focus was twofold: extensive exploration of the Colorado Plateau and surrounding areas, and tapping into the already existing, albeit limited domestic uranium sources, primarily derived from old radium and vanadium mines. The mandate was clear: acquire as much uranium as possible, as rapidly as possible, and as cost-effectively as possible.

In an effort to revive uranium production, the AEC entered into contracts with companies such as US Vanadium (USVC) and the Vanadium Corporation of America (VCA).[23] This was not the first time that they had been involved in the search for uranium, with both having entered into covert contracts with the government during the Second World War, supplying them with the tailings from vanadium processing, which contained uranium. Vanadium, again back in high demand during wartime, played a pivotal role in the secretive procurement of uranium.

With the end of the war, the market for vanadium ore took a sharp nosedive, experiencing a staggering 72 per cent decline between 1945 and 1946.[24] With the advent of the new post-war government agreements, companies swiftly ramped up production once again. Starting from 1949, VCA, in particular, took decisive measures to reopen their vanadium mines and set up two treatment plants.[25]

The AEC also entered into a contract with the US Geological Survey (USGS), a long-established agency in the Department of the Interior.[26]

Under the agreement, the USGS was tasked with carrying out a comprehensive survey that involved both field and laboratory investigations. Their mission was the 'development of a full knowledge of the geological occurrence of radioactive materials and toward an appraisal of the resources and availability of such minerals, ores, and other raw materials within the United States, its territories and possessions.'[27] They deployed over a hundred geologists to assist in the search for uranium and to further map out the geological landscape to support the expansion of the domestic uranium industry.[28]

Between 1948 and 1956, the USGS and AEC invested significant resources, amounting to millions of dollars, in drilling exploratory holes near potential uranium deposits.[29] The exploration efforts included a wide range of methods, such as large-scale diamond drilling, establishing exploration camps and airborne reconnaissance.[30]

This last method was conducted using a fleet of light planes and a squadron of Piper Super Cub planes.[31] These aircraft were equipped with dangling scintillometers, which were sophisticated radiation detection devices suspended underneath with wires.[32] These instruments were capable of picking up ground radiation readings up to 1,500 feet away.[33] However, before conducting these flights, the counters on the planes' instrument dials had to be shielded, as the glow in the radium paint used could interfere with the readings and render them useless for detecting uranium.[34]

The daring bush pilots tasked with this reconnaissance would skim along the ragged canyon rims at speeds of about 55 miles per hour, flying as close as 30 feet from the edges. The pilots would meticulously scan the landscape for anomalies, which were areas with radiation levels higher than the background count. These findings would then be recorded on maps, indicating potential uranium-bearing formations, for future investigation.[35]

To encourage the industry to develop by April 1948, the AEC launched a series of incentives, issued in the form of five numbered circulars, to activate the domestic uranium industry by bringing independent miners into the cause.

In effect these established guaranteed prices of 'domestic refined uranium, high grade-uranium bearing ores and mechanical concentrates'. Circular Two offered a bonus of $10,000 for the discovery and production of high-grade ores, Circulars Three, Four and Five offered hauling allowances and developmental loans.[36] As the programme expanded, the emphasis also shifted from a focus on high-grade ores to encompassing lower-grade ores more commonly found on the plateau, like carnotite.[37]

With this lucrative encouragement came thousands of dedicated uranium prospectors tirelessly scouring the lands for the coveted minerals. Among them were seasoned professional geologists, but there were also countless others, an estimated 10,000 at least, who pursued their passion for uranium hunting during weekends and in their spare time. As Jesse C. Johnson, Director of the AEC's Raw Materials Divisions, put it:

> Anyone can join the hunt for uranium. Indeed, among the countless thousands of uranium prospectors in the field today there are women and children, prospecting right alongside the family breadwinner. Some are at it full time. Others part time, on weekends and holidays, even at night. Some combine it with travel and vacation fun. Others make it a hobby.[38]

These so-called 'weekend prospectors' were also an important part of the carefully controlled AEC narrative, which portrayed the uranium boom on the Colorado Plateau as a spontaneous tale of rags to riches American go-getting rather than a cynical exploitation of the land.[39]

As droves of enthusiastic newcomers joined the burgeoning field of uranium prospecting, a whole new industry sprouted to cater to their needs. One of the biggest segments of this industry was focused on providing essential tools and equipment, as well as specialised clothing. *Life* magazine provided cost estimates for the new uranium prospector, with a basic kit estimated to set you back around $180. There was also a more expensive kit, with a correspondingly larger price point, that included a four-wheel-drive jeep, knapsack, scintillation counter, Geiger counter, wooden carrying case, ore pick, portable drawing board, hundred-foot tape for measuring claims, claim notices and a coffee pot.

The article also featured a fashion spread that has since become a source of fascination, particularly in the world of social media, where it is frequently shared with an aura of disbelief.[40] The family in the photographs are depicted against a fake backdrop of mountains and cactuses, dressed in what *Life* referred to as 'prospecting duds'. The father is seen sporting a stylish black outfit with red breast pockets, collar and a cap, exuding a sense of sartorial flair (well, at least a very 1950s look). The mother takes centre stage in the feature, clad in a bright orange jumpsuit referred to as the 'U-235', complete with perfectly coiffed hair and a shovel in hand. The daughter dons a smaller version of the mother's outfit, known as the 'Diggerette Jr', complete with black detailing on the pockets.[41]

While the spread is joyful to look at, it wasn't exactly what the majority of prospectors would be wearing – nor would it even have been advisable to do so. They were just not practical, as contemporary uranium hound Al Look pointed out in his book *U-boom*:

> In front of moth eaten scenery on a black canvas ground the *Life* teenager in braids wore a bright red 'Diggerette Jr.' suit and cowboy boots that were made only for riding horses. It was a good outfit for digging clams at Coney Island.
>
> This little lady carried a quart canteen, evidently the family water for a day's work. The marcelled wife lugged a shovel heavy enough to drag a pack mule, and a geologist's pick,

which was all right, except the horsehide gloves, which she didn't have, are recommended to handle the rock picked out of the cliff. Her form-fitting suit, called the 'U-235' model, had patch pockets as big as saddle bags, to flop on the back of each leg and be torn off by the first sage bush. She had clodhopper shoes to make heel blisters, shoes loose enough to catch all the migrating gravel on the mountain. The male wore earphones and used both hands to carry a scintillator and probe, leaving nothing but feet for climbing. He was posed like an African big game hunter with a red shoe lifted on a papier-mâché rock. They all wore headgear that would shed water like a sponge.[42]

Another piece of essential equipment available at the time was the Geiger counter, which first went on sale commercially in the late 1920s. There were plenty of adverts, publications and catalogues offering Geiger counters and, crucially, explaining how they worked:

> In short, the Geiger counter is like a radio. Your radio is a power unit designed to pick up sound waves in the air and translate them into the original sounds, voice, music or just plain noise. The Geiger counter is a power unit designed to detect radioactivity in the air and tell you, via different mechanical means – such as meters, clicking sounds or flashing light tubes – that a certain amount of radioactivity has been detected.[43]

Or, as even more simply described by Lucille Ball in a 1958 episode of The Lucy–Desi Comedy Hour, 'you turn it on like this. Then you hold this up to your ear ... and then when it gets near uranium it starts to click'. It is a good piece of scientific communication in an otherwise madcap (is there any other kind!) episode where Lucy convinces the gang – and guest star Fred

MacMurray – to go uranium hunting in the desert outside of Las Vegas.

The popularity of Geiger counters was such that in 1953 it has been estimated that 35,000 were sold by mail order alone.[44] Taking advantage of this popularity was the Radiac Company, who opened 'America's first Atomic age department store'[45] for prospecting equipment on the seventh floor at 489 Fifth Avenue, New York City. They stocked many types of Geiger counter, such as the Lucky Strike, selling for $99.50, and the Ferret, which at $20 was the cheapest and smallest they offered.

If you enjoyed making your own technology, then there were plenty of publications offering detailed instructions on how to go about that. *Popular Mechanics* magazine, for instance, featured an advert for a kit that claimed: 'In one evening, you can assemble a VACATION GEIGER COUNTER.'[46]

Once kitted out in a U-235 suit with a Geiger counter in hand, there were plenty of publications aimed to show the amateur the practicalities of uranium hunting.

The AEC and the Geological Survey had joined forces to publish a pocket-sized manual titled *Prospecting for Uranium*. Priced at just 30 cents, this little booklet became a sensation, flying off the shelves and becoming a bestseller in no time. In the first publication year alone, over 16,000 copies were sold, and by 1950, nearly 70,000 copies found eager buyers.[47] Such was the demand that the AEC had to order an extra 6,000 copies for distribution, and the government printing office struggled to keep up with the overwhelming requests.[48]

Encouraged by the success of the AEC publication there were hundreds of other uranium-hunting handbooks released of varying usefulness, including *Uranium Prospecting: A Complete Manual*, *The ABC's of Uranium Prospecting: A Guidebook for the Amateur*, *Uranium Prospecting Handbook*, and *From Rags to Riches with Uranium*.

These manuals covered a wide range of topics, including how to identify potential uranium deposits, the intricacies of staking a claim, drilling techniques to determine deposit size, forming corporations, selling stock for mine development and locating AEC ore-buying stations.

As well as the numerous books there were magazine articles, with headlines such as 'How to Hunt for Uranium' (*Popular Science*), 'Out Where the Click is Louder' (*Time*) and 'History's Greatest Metal Hunt' (*Life*) to advise the casual prospector.[49] There were also plenty of advertisements for the numerous schools, courses and mobile schools aimed at teaching people how to identify and mine for uranium. These courses gained popularity, with some attracting as many as a thousand prospective students.[50]

In fact, some unconventional locations were used for field work and practical training during these courses. For instance, Mesa College in Grand Junction offered its first program in prospecting in 1955, and students reportedly conducted their field work at the city swimming pool, the concrete of which was made of radioactive sand hauled in from the uranium-mill dump located south of the town, so offered the perfect way to test Geiger counters.[51]

If you wanted a low risk, low effort, prospecting experience, then board games were available. *The Game of Life*, where a player's success throughout their lifetime is measured by money and achievement, featured a prize square which read: 'DISCOVER URANIUM! COLLECT $240,000.'[52] Other games like the simply titled *Uranium* or *Uranium Rush* also had the goal of striking it rich, but it wasn't always a settled matter – like prospecting in reality – whether you would make any money. In *Uranium* spinning a wheel would determine your fortune, while in *Uranium Rush* a small Geiger counter included in the game could be the passport to a small fortune.

The publicity surrounding all of these activities fuelled a wide-spread desire to become a uranium prospector; the AEC's Division of Raw Materials office fielded more than 12,000 letters.[53] Some of the correspondences were more serious than others, and show some mixed levels of understanding about the effects and properties of radioactivity. For example, one letter included a story from a lady who believed that her pasture might contain a uranium deposit because her cows grazing there were losing their hair. Another man claimed to have struck it rich because his watch always stopped when he passed a certain location. Yet another elderly man was convinced he had found uranium because his ears began to ring when he explored an abandoned mine.[54]

As Muriel Mathez, chief minerologist for the Raw Materials Operation Laboratory, told the *New Yorker*:

> We've had twenty five hundred samples of rock sent in to us since the government uranium-hunting program began last year ... and the number of samples we get per day keeps going up ... Ninety per cent of the samples we receive never get past our first Geiger counter test, for the simple reason that they're not radioactive. Heaven knows why people send them to us. We've had all kinds of things turn up in the mail, including two bricks.[55]

The letter writers and the sample senders were responding to a mythology around uranium hunting – that it was possible to become a 'uraniumaire'. Uranium was rapidly becoming the twentieth century's gold rush – and for many the dream was to find a source and become rich beyond your wildest dreams or, as sung by Frank Sinatra and Celeste Home in the song 'Who Wants to be a Millionaire' from the movie *High Society* – to 'Have Uranium to Spare'.

The first of the successful prospectors was Navajo Patricio 'Paddy' Martinez, who in the spring of 1950 discovered uranium

in the form of carnotite, near Grants, New Mexico. He received a
$250 finder's fee, as the claim turned out to be on the property
of the Santa Fe railroad and he wasn't able to exploit the find
himself. It would be the first, but certainly not the last find in
the area.[56]

Whereas there was plenty of press attention for Martinez's
find, the story of Charlie Steen and his wife Minnie Lee (M.L.)
was one of the most enduring legends in the field of uranium
prospecting. The couple endured numerous hardships and years
of searching for uranium, which took a toll on them and their
children. They borrowed money from M.L.'s mother and friends
to fund their prospecting efforts but kept running out. Steen
staked out a dozen claims in and around Moab, Utah, and gave
them evocative Spanish names like *Mi Corazon* (My Heart), *Linda
Mujer* (Pretty Woman), *Te Quiero* (I Love You) and *Mi Vida* (My
Life).[57] With the ore samples in hand, they sought backing to
develop their claim and sought publicity to attract investors. A
friend eventually gave them $20,000 to form a company called
Utex Exploration Company, and they began digging a small
shaft.[58] In December 1952, the couple's perseverance paid off
when they hit a fourteen-foot lode of pitchblende, a rich grade
of ore containing upwards of 70 per cent uranium.[59]

According to legend, after receiving confirmation through
a Geiger counter that the ore they had discovered was indeed
radioactive, Charlie ran excitedly to his home to share the news
with M.L. 'We've hit it!' he said. Her response was: 'Well, dar-
ling … I always knew you would.'[60]

The Steens were much in demand with publications like
Newsweek, Time, Business Week and *Women's Home Companion*,
which published in-depth profiles. And the ongoing saga of the
Steens' adventure captivated the public. By 1955, their mines
had amassed a staggering value of over $130 million, as they
shipped 200 tons of exceptionally pure uranium to US Vanadium

for processing on a daily basis. Soon after, the Utex Company made the decision to establish its own independent processing mill in Moab.[61]

Charlie, in particular, was renowned for his flamboyant personality and love for publicity and extravagance, earning him the moniker of the 'Uranium King'. He spared no expense in indulging in a lavish lifestyle, hosting opulent parties replete with champagne and gourmet delicacies flown in specially for the occasion.[62] There were stories of chartering private planes to attend dance classes in nearby Salt Lake City and hosting celebrity dinners. He even had his old, worn-out boots bronzed and displayed on his mantlepiece as a reminder of the past difficulties he had faced, 'to remind me, in case I forgot, how "easy" it is to earn a million bucks'.[63]

It wasn't only the Steens who were profiled in magazines and newspapers. Newspapers covered the stories of these individuals who had willingly abandoned the comforts of their previous lives in pursuit of the life of a prospector. This theme of leaving behind the 'easy life' for the allure of prospecting, with its inherent sense of adventure and romance, was a popular topic.

It is from these sources that we know of Stella Dysart, a successful businesswoman in LA, who made a bold move by purchasing a sheep range north-west of Grants and dedicating 25 years to drilling for oil and then, when that didn't pay off – uranium. Her determination was rewarded when she discovered a uranium seam on her land.[64] Another, Jeanette Martin, known as the 'Prospecting Grandma', was offered a staggering $1 million for her uranium find in 1955.[65]

The excitement surrounding uranium was not limited to individuals, either – entire communities were swept up in what some called 'uranium fever'. Take the towns of Moab, and Grants. These places were transformed by major uranium strikes and were quick to proclaim themselves as the 'Uranium Capital of the World'.[66] In fact, the two towns even vied for the title.[67]

Moab largely owed its uranium notoriety to Steen's massive discovery. As people flocked to the area eager to emulate his success, new infrastructure had to be built to accommodate them. Trailer courts sprang up, and essential services such as water and electricity were extended to these new developments. New retail centres were established, and roads were rerouted to cope with the increased traffic.[68]

In the midst of all this activity, the community developed an identity around their uranium rush. They had an Atomic Café and a Uranium Office Building. The rodeo, once known as the Red Rock Roundup, was rebranded as the Uranium Days Rodeo. The town even boasted a Uranium Club, where the sign in the window read 'No Talk Under $1,000,000'.[69]

Just over 300 miles away, Paddy Martinez's uranium discovery in 1952 had changed the town of Grants completely. Almost overnight, it experienced a population boom as it transformed from the 'Carrot Capital of the World' to the 'Uranium Capital of the World'. The first uranium processing mill in the area was built in 1953, with a second one following in 1955. By the end of 1958, four more mills had been built.[70]

Like Moab, Grants had its own share of uranium-themed businesses and events. One such business was the Uranium Café, which opened in 1956 and served up the 'Finest American, Chinese and Mexican Dishes'. Located in the heart of town along Route 66, the café's sign even evoked a mushroom cloud.[71]

Grants also held citywide festivals to celebrate its newfound identity. In 1956, a local miner was named 'Uranium Prospector of the Year' during one such celebration. The previous year, there had even been a 'Uranium Queen' contest, with the winner receiving a whopping ten tons of ore. The 'Miss Atomic Energy' pageant was also held in the town, with another truckload of uranium ore as the prize. This event was sponsored by the Uranium Ore Producers Association and the Grand Junction Chamber of Commerce.[72]

Other places in the Four Corners region, with less obvious uranium links, also took part in the craze. In July 1954, *Life* magazine memorably captured some of the 'silly sidelights' around the uranium boom in Salt Lake City, which: 'include a uranium-burger which is really just a nonradioactive hamburger. Another shop offered the uranium sundae, a concoction of ice cream and sherbet smothered with whipped cream, pineapple and jelly beans.'[73] The photographs accompanying this five-page spread tell us that the burger cost 45 cents and the ice-cream sundae 35 cents.

Life also introduced the reader to the penny stock boom, the epicentre of which was Salt Lake City, which for a time became known as the 'Wall Street of Uranium'.[74] This all seems to have started when local man Jay Walters purchased some long-shot claims and established the Uranium Oil and Trading company, offering 3,000,000 one-cent shares. When traditional brokerage firms refused to deal with him, he turned to Jack Coombs, who opened a sales office on a coffee counter at the Continental Bank & Trust Company, run by Frank Whitney. Here, they began selling shares of the company at one cent each, and allegedly sold $10,000 worth within a week.

The practice of selling shares in mining companies, whether they were established companies branching out into uranium prospecting or new companies with unexplored claims, was a common means of raising funds, with investors receiving a share certificate as proof of ownership. As Harry Kursh explained in the popular manual 'How to prospect for uranium':

> When you buy uranium stock you're investing in a company that in some way deals with uranium. Sometimes it is a

company that is just being organized to conduct prospecting
and is using your money, your investment, for that purpose.
Sometimes it is a company that has already staked out some
claims but needs money to buy equipment to explore the
claims. Staking a claim is one thing. Finding out how much
uranium the claim contains is another. It takes exploration to
find out. Exploration could be anything from chopping rocks
to drilling holes with costly diamond drill rigs. Other times, it
may be a company that already has explored its claims, knows
pretty much what to expect but needs money with which to
buy equipment and hire men to actually mine the ore and sell
it to the government.[75]

The shares were sold by brokers, which was essentially anyone
that had bought a brokerage licence.[76]

The business concept gained immense popularity, leading
Frank Whitney to shut down his coffee counter and team up
with his brother Richard to establish Whitney Investments
Company. Coombs also followed suit. Walters went on to establish Aladdin Uranium and sold even more shares, soon other
small prospectors and mining companies joined the fray.[77] A
directory from 1954 listed over 500 uranium corporations in
the Colorado Plateau region, with Salt Lake City as the centre,
before spreading out through stock-trading centres across the
nation.[78]

In May 1954, trading in uranium mining company shares
peaked at a staggering 7 million per day.[79] However, it eventually
simmered down to what one overwhelmed broker dubbed a 'dull
roar' of a daily 5.5 million shares sold.[80] One broker said: 'Las
Vegas isn't in it any more. Salt Lake is the gambling capital of
America and we're doing it with uranium.'[81]

It is estimated that hundreds of thousands of Americans
participated in this frenzy. These included individuals from

different economic backgrounds, such as housewives, doctors, businessmen, teachers, bank officers and cab drivers.[82]
Like many of these types of schemes, those who got in early
reaped the greatest financial rewards. For the fortunate investors who bought into companies like Lisbon Uranium, the
value of their investments skyrocketed from 20 cents to $3.12
per share in just six months.[83] These success stories were
enough to maintain high levels of interest and enthusiasm
among investors, even if other companies weren't perform-
ing as well.[84] As Kursh again explained to his contemporary
readers:

> You can make money in uranium stocks – if you are careful and
> if you're lucky. You have to be careful in the stocks you select,
> making sure you're not dealing with swindlers or with brokers
> who are here today, gone tomorrow. You have to be lucky, just
> like the prospector in the field, sweating out the day when the
> company you invest in makes a strike that brings in a bonanza.
> If you get taken in by a phony – and nobody knows how many
> there are – luck won't have anything to do with it. You can kiss
> your money good-bye.[85]

As the uranium investment frenzy continued, it became increasingly difficult to earn any substantial profits, and share values
inevitably began to decline. The *Life* magazine profile of Salt
Lake City reported that share certificates were being handed
out in a number of jocular ways, including a drug store contest
that would give a thousand shares each week for ten weeks to
the person who could catch the largest fish.[86] And there were
increasingly reports of swindlers:

> One of the nation's biggest uranium penny stock underwrit-
> ers, Walter T Tellier of Jersey City and Englewood, N.J, has

been indicated for the third time on charges of mail fraud, federal securities fraud, and conspiracy ... through Tellier's stock manipulations more than 30,000 penny stock investors throughout the country have been swindled out of at least $5 million ... Tellier is awaiting trial on two previous indictments – one charging him and two Seattle, Wash. businessmen with the nearly $900,000 fraudulent sale of Alasca Telephone Company bonds to nearly 1,400 persons, and the other accusing him personally with defrauding some 50,000 persons out of about $15 million through a fraudulent sale of uranium stock.[87]

Investors who bought shares in fraudulent or failed uranium companies were left with worthless certificates. However, they were still better off than those who fell for the scam of buying stock in a uranium mine on the moon, which was reported in *Popular Mechanics*. The magazine issued a warning to readers to 'Beware of Lunar Uranium Stock'.[88]

And it wasn't just in the stock market. Criminality, from claim jumping to murder, was rife within uranium prospecting as well. There were 'uranium murders' which were sensationalised by the press, as in the case of LeRoy Wilson whose body was reported to have been found with '.45 slugs, a piece of hot uranium ore in one hand and a clicking Geiger counter in the other'.[89] Other crimes that didn't make the press were the many cases of Indigenous Americans, like Paddy Martinez, who helped to locate uranium deposits but were often cheated out of fair compensation.[90]

In fact, very few prospectors actually made much money, and even fewer collected the bonus offered by the AEC, which, according to a contemporary author and uranium miner, had requirements that 'resembled those of an insurance policy which pays off only if you are stricken by lightning at high noon of

July 4 while riding in a 1962 Chrysler on the road from San Jose to Los Gatos, California'.[91]

Befitting such a rollercoaster industry there was a plethora of country and western songs that were released in the mid-1950s about the uranium rush, with titles like 'Uranium Blues', 'Uranium Rock' and 'Uranium Fever'. From these we hear about miners who rise early to 'dig that yellow stuff that makes the atom bomb' or divorce caused by catching uranium fever. Some of these songs were based on real-life experiences. Riley Walker and His Rockin-R-Rangers recorded 'Uranium Miner's Boogie', and – giving credibility to his tale – Walker actually had been a miner himself.[92] Singer Elton Britt and his wife Penny – who also gets songwriting credit – had gone to Utah together to find uranium and based their song 'Uranium Fever' on the experience.[93]

For every Steen, Dysar and Martin, there were many thousands who were not successful. George Polites, a cab driver in Miami, abandoned his former life to become a prospector in New Mexico for six months in 1955.[94] He spent $9,000 of his own savings, staked out 128 claims and came away with very little.[95] Ed Walls was profiled in a piece called 'Uranium Let Him Down (To Success)' in *Popular Mechanics* magazine, with the story that he sold everything, took his family to Utah when the 'uranium bug hit him a couple of years ago' and lost it all. It all turned out fine because he gave up, bought a rug machine and made his fortune that way.[96]

And the tale of disappointment in uranium hunting wasn't just on an individual basis – whole areas could also be thwarted. The whole town of Grants had caught uranium fever on 23 April 1953 as Geiger counters registered high readings in the area. However, after careful investigation, it was found that the increased radiation levels were likely because of an atomic test conducted upwind in Nevada, much to the disappointment of many in the population.

It was not uncommon for companies to warn new prospectors about getting overly excited about such spikes, as they could often be attributed to external sources. The Uranium Engineering Co put up a sign:

Wait a Minute – have you considered 'Fall-Out?' The current atomic tests at the AEC proving ground in Nevada will result in a heavy fall-out of radioactive particles. This area will doubtless experience at least a portion of the fall-out. The drastic increase in background count means that counters and scintillators will read abnormally high. Before you decide your instrument isn't working properly take the fall-out into consideration.

While disappointment seemed to be part of the prospector's life in 1952, both *Time* and *Life* magazines provided readers with a glimpse into another way of commercialising uranium with articles on the Free Enterprise Mine and its dynamic owner, Wade V. Lewis. While both tell his story, it was *Life* magazine's introduction that truly captured the essence of Lewis and his venture, setting the tone for the rest of the article: 'Near the little Montana mining town of Boulder last week a couple of second-rate uranium mines were busily marketing a new and profitable stock in trade. Their commodity: hope.'[97]

Wade V. Lewis, J.T. Lewis, and Walter B. Smith had previously founded the Elkhorn Mining Company. As equal shareholders and original directors, they set out to fulfil their mission 'to search, prospect and explore for ore and minerals of any kind'.[98] Such efforts proved fruitful, and they discovered uranium ore in 1949. For two years thereafter, this ore was sent to the Vitro Chemical Company in Salt Lake City for processing.[99]

For the next part of the story we have to turn to company legend – a story which has a husband and wife from Los Angeles stopping off at the mine in 1951. During their visit, a joke was made about the potential of the radioactivity to cure the woman's bursitis. Well, wouldn't you know it, the joke came true, and the next day, the husband called to confirm. News of the miraculous recovery spread by word of mouth, and soon the Elkhorn mine began receiving more health visitors.[100] So many people came that Wade Lewis was forced to make a decision: 'Visitors were welcome but finally interfered with the primary uranium ore and mine development to such an extent that it became necessary to either deny entrance to the mine or provide proper facilities for visiting.'[101]

Realising the potential for a lucrative new venture, the owners of the Elkhorn Mining Company quickly acted to accommodate the influx of visitors. They transformed the space into a health centre, which became known as the Free Enterprise Uranium Mine.[102] Prior to opening, they invested in additional facilities such as lighting and an 'Automatic Otis 12-passenger elevator' to transport visitors down into the mine, at a cost of $50,000.[103] On 23 June 1952, they officially opened for business, and began processing visitors, as a leaflet in the archives of the American Medical Association explains:[104]

> Those people visiting the mine are first documented at the company office, Boulder Bank Building, Boulder. At the mine, each person descends from the surface to the 85-foot mine level and is assigned one of the underground stations, numbered 1 to 35. Special rooms have been excavated and timbered to accommodate wheel-chair and litter cases ... A registered nurse is in attendance in the waiting room or underground. This nurse interviews each visitor each day, noting signs, symptoms and improvements.[105]

The profiles in both *Time* and *Life* early the following month helped to give the venture national publicity. A powerful black and white image in *Life* documenting people at the health spa is captioned: 'patient sufferers sit quietly but reveal in their faces the uncertain hope they feel in the presence of the invisible and mysterious gas which they think may somehow help them. During one trip a woman grew hysterical and was carried out screaming, "Oh God, I can feel it! I can feel it!"'[106]

While the overall tone of the coverage in *Life* seems rather tongue in cheek, *Time* was more brutal in its commentary:

> There is nothing in medical or nuclear science to suggest that the tiny, harmless doses of radiation absorbed by visitors can affect joint diseases. Nevertheless, a lot of people who go down the mines feel better afterward. It is also true that some people with pains in their joints (from either disease or a state of mind) feel better after carrying potatoes in their pockets. There is no reason to believe that sitting in a mine is any less effective for such cases than carrying a potato. But in Boulder and Basin it is more expensive.[107]

Despite this rather negative publicity, Lewis's venture was a success, with visitors being charged $10 per visit to sit in converted mine shafts and inhale. What they were breathing was explained in the brochure: 'It's Radon Gas – NOT Uranium Ore! Every Uranium Mine does not yield, replace and maintain daily a CONSTANT SUPPLY OF RADON GAS as does the FREE ENTERPRISE MINE. RADON GAS, not Uranium Ore, is the SOURCE of the EFFECTIVE RADIOACTIVE ELEMENTS!'[108]

Radon gas had continued to be a popular health treatment since its beginnings in Jáchymov. The success of it there had expanded radon spa treatments to many places throughout the world. In Europe towns like Carlsbad, Vichy, Hamburg, Bath and Baden-Baden, were used as a model for health resorts throughout the United States. Radon and radium treatments were available at Saratoga Springs, which promoted its radioactive waters as a treatment for gout and to lower blood pressure, and lauded its ability to act with 'a favourable effect upon organic metabolism'.[109] At Hot Springs, Arkansas, you could get 'Hot water from Spring to Tub without exposure to Air, thus retaining all Radium Gases' in the Superior Baths or by drinking it from one of the public Hot Water Drinking Fountains dotted around the area.[110]

And while by the 1950s the worldwide popularity of radon treatments had rather waned, there were still plenty of places offering this style of treatment. Wade Lewis's inspiration in setting up the Free Enterprise was the Heilstollen, or thermal tunnel, just outside of the spa town Bad Gastein in Austria.

Prior to this, the site had been a gold mine. During the Second World War, the Nazis had taken an interest in the area and had excavated a tunnel into the mountains in search of gold. Although they did not find any remaining, the chief mining engineer, Karl Zschocke, discovered small amounts of secondary uranium minerals. On top of that, the mine workers, who were primarily prisoners of war or conscripts, were noted to be in good health, considering the physical nature and inherent dangers in their work, and indeed did not seem to suffer from colds during the winter.

As the war was coming to an end and the hoped-for gold had not materialised, Zschocke received orders to remove all the equipment and destroy the tunnel. But he deliberately delayed the process and managed to postpone it until the end of the war. This meant that the artificial ventilation systems and train rails were still in place. The Austrian Academy of Sciences investigated the air shafts and confirmed that the tunnels had concentrations of radon. A hot-air emanatorium, the Heilstollen, opened in 1952. It was said that the tunnel's combination of radon gas, high humidity and temperatures would be beneficial to health.

And since then, health-seekers have been funnelled between the rest-cure stations by an electric subterranean railway. If you visit today – and there are thousands that do – you can still experience the two-hour treatment session encompassing four areas with different air temperatures, ranging from 37 to 41 degrees Celsius at the highest.

While the Free Enterprise Mine was probably the first of its kind in the United States, it was certainly not the last. Shortly after its opening, the Merry Widow Mine, located nearby, began admitting visitors on a voluntary donation basis. Other abandoned or low-production mines soon followed suit, particularly along Highway 91 on the outskirts of Boulder.

Mines were not the only way to get the health benefits of uranium – there were other options, including tunnels, 'sitting houses' and uranatoriums. Jesse Reese, a dairy farmer from Newburg, Texas, was perhaps the first person to set up one of these standalone treatment centres. The story goes that in 1953 he had a bad leg that miraculously improved after he discovered that his farm was slightly radioactive due to the presence of uranium.[111] Word of his apparent cure quickly spread, and people started flocking to try the treatment themselves. By 1955, Reese had converted his garage and dairy barn into a 'sitting house' that accommodated 112 people, each paying $2 for 90 minutes. They even had a restaurant with a fairly extensive menu including sandwiches, cakes, hamburgers, pies, cold drinks and coffee. On one of his busiest days, Reese claimed to have welcomed 461 people.[112] There was a flurry of publicity, and Reese even appeared on the *Home* TV show hosted by Arlene Francis in New York.[113]

After this success, reports began to circulate about the 'Uranium Trail' in Texas. This trail consisted of 25 different 'sitting houses'. These establishments had extended operating hours, usually from 9am to 10.30pm, but some remained open 24/7.[114]

The sitting houses on the trail varied greatly in size and building type, ranging from private homes to barns and garages. Some were described as being quite impressive, with amenities like snack bars, cold-drink machines, restaurants and sofas.

The treatment methods varied between businesses. Jesse Reese had a chair filled with uranium dirt. Some places had benches filled with soil, while others had shallow troughs where customers buried their bare feet in the dirt.[115] Certain establishments offered complete immersion while lying down. At Frank and Annie's, customers sat with sacks filled with uranium sand on their heads and shoulders, and their feet in troughs while their hands were placed in cardboard boxes filled with sand.[116] Some even lay prone and requested to be covered with sand, thinking that the more body surface was covered, the more likely the treatment would work.[117]

The owners of the sitting houses were largely cautious with the language they used to describe the experience and the likely outcomes. This was important, as they needed to avoid breaking any rules around providing medical treatments without licences. By calling them sitting houses, the owners could claim they were simply allowing people to pay to sit there, getting round the regulations and restrictions.

Typically, any responses to questions from the press would lead to noncommittal answers like 'people who have used the sitting house reported the sittings made them feel better'.[118] Although Mrs Thornton of the Uranium Sitting House, Albuquerque, went a little further:

> We don't know what the sand does for people. But it must do something. People come here with a condition of the back or something and lie in the sand an hour and a half. They say they really feel better when they go home. Maybe it's not so much the uranium that helps. But there are 12 minerals the body needs in the sand, and the uranium makes them ready to go into the body.[119]

Very few of these places had actual access to uranium on their site; instead, they bought it from uranium mining companies. George Polites, the Miami taxi driver turned unsuccessful prospector, skipped the middleman altogether and used the uranium from his low-grade mines to start the Uranium Health Center in Miami.[120] Polites offered health seekers the opportunity to lie in a long wooden box filled with low-grade uranium, available in three colours – brown, grey and yellow. The cost of the treatment was $60 for six sessions, which had to be paid in advance, and each one lasted for 90 minutes.[121]

The Merry Widow Mine also sent out uranium ore as a product to a number of commercial organisations.[122] Blair Burwell, referred to in the popular press as 'Mr Uranium', was manager of the Climax Uranium Company, which sold its tailings to manufacturers for use in rheumatism pads.[123]

There were also many products that used uranium ore, such as the Torbena Radioactive Water Jar, which was lined with torbernite.[124] Another example is the Revigorette, a small Bakelite jar lined with uranium ore which was patented and trademarked for: 'Containers used for substances to which it is desirable to impart the medical property of radioactive emanation.'[125] In the case of the Revigorette the intension was to impart 'beautifying radioactivity to every face cream'.

Rheumatism pads were also a popular market for uranium tailings. These pads were essentially fabric bags of different sizes that were filled with some sort of ore. The idea was that they were placed on the part of the body that hurt and would heal through their radioactive properties. Several brands were introduced to the market. The Wonderpad range included Wondergloves and Wonderpad boots 'for lounging or sleeping; to name but a few'.[126] The Gra-Maze was marketed as a 'personal radioactive uranium comforter – your own health mine in miniature'.[127] The Cosmos Radioactive Pad came with the directions: 'Do Not Open. Not to Be Taken Internally. Place Pad Under Pillow or Mattress.'[128]

While the American Medical Association (AMA) had been strong advocates for radium therapy in the early-twentieth century, they had eventually changed their stance, and by 1933, as the dangers around the substance became apparent, the association banned radium as an internal medicine.[129] By the 1950s, the AMA had also denounced radon therapy as a medical hoax, labelling its advocates as quacks and charlatans.

There were so many of these style of uranium products that the AMA and the authorities struggled to contain and control this new form of medical quackery. In their archives are hundreds of enquiries from members of the public, physicians, the Better Business Bureau and newspapers trying to find out more about these uranium treatments.

The AMA tested the products that were referred to them, either with a Geiger counter or in the laboratory, and then shared the results in their publications or in response to inquiries. Sitting houses and tunnels were dismissed with a brutal quip: 'Fascination for the gullible is only known effect of uranium tunnels.'[130] The Wonderpad was deemed 'worthless' and a 'fake medicine pitch' that 'did not contain uranium or any other radioactive material but was filled with chunks of rock and pulverized dirt'.[131]

The AMA even contacted the AEC to see if these companies and sitting-house owners were breaking licensing regulations around uranium which was technically under their control. In a letter dated 14 October 1952 the AEC confirmed: 'Unfortunately, in accordance with the Atomic Energy Act, we have no control over the disposal of radioactive material containing uranium in the following amounts: (a) Materials containing less than .05 per cent of uranium or thorium, or any combination thereof are not considered "radioactive source material" and are not subject to our licensing regulations.'[132]

While that technique hadn't worked, wherever possible the AMA took steps to refer suspicious products to federal organisations, such as the Post Office fraud division for products shipped by mail, or the US Food and Drug Administration (FDA) if they believed it was misbranded. As the radioactivity was deemed too low to be a health hazard, this misbranding tactic was more effective in successfully stopping the sale of these products.[133] If an item was deemed as misbranded, i.e. making a claim or implying that it was an effective treatment then, under the Food, Drug and Cosmetic Act (1938), the FDA could seize and destroy them or close down the service.

George Polites' Uranium Health Center in Miami was forced to shut down in October 1956 after their uranium ore was seized for misbranding.[134] This led to a lengthy battle with federal authorities, the Dade County Medical Association and various critics, which played out in the press. The confiscated ore was stored in an international bonded warehouse before eventually being destroyed. The Uranium Health Center's publicity director, Sam Wallace, rather churlishly stated that they were planning to close anyway, and that the seizure saved them from having to incur moving expenses.[135]

Olen and Jewell Cates, who ran the Uranium Sitting House, just outside of Odessa, Texas, took aim at state officials, saying they 'should realize that all the basic therapeutic applications in use today by doctors all over the world sprang originally from the earth in one form or another. They were placed there by the good Lord to be discovered by man for the benefit of mankind.'[136]

Medical treatments using uranium, radium or radon may have been officially relegated to being mere quack cures, but the mainstream medical use of artificially created radioactive isotopes was only increasing in this period.

At the newly created Oak Ridge National Laboratory, which became known in the press as 'the atomic drugstore', a

comprehensive programme was established to provide radioisotope for research, therapy and industry. This was an ideal place,
as the existing reactor and chemical separation plant, along with
the skilled personnel, could be used.[137]

The facility was described in contemporary publications as
partitioned with a concrete wall measuring six-feet high by two
feet in depth. Above the wall was a tilted mirror, which reflected
the room behind it. And within that space were large concrete
chests with metal drawers. Remote-controlled robot hands were
used to lift the bottles from the drawers, insert a glass pipette,
remove 'a thimbleful of radioactive liquid', which had been generated by the nuclear reactor, and then place it in a sealed lead
container.[138]

As an employee from Oak Ridge assured F. Barrows Colton,
Assistant Editor of the *National Geographic* magazine in 1954:

> We keep our isotopes behind that barricade, handle them only
> by remote control and look at them only indirectly, in the mir-
> ror because the concentrated radiation from them otherwise
> would be dangerous. The small amounts we send out to our
> 'customers', however, are quite safe when properly handled.
> We're learning to live with atomic energy, just as we do with
> fire and electricity. We do not fear it, but we respect it. Watch
> and I'll show you how we fill and order.[139]

Their eventual isotope programme was officially inaugurated in
August at a carefully planned ceremony.

The Director of ORNL, Eugene Wigner, asserted: 'We hope …
that this day will mark a turning point in the history of atomic
energy and that which has been used so effectively for the purpose of destruction will henceforth be used with even greater
efficiency for the saving of lives, and the increase of our knowledge for the benefit of mankind.'[140]

During the 1950s, interest in radioisotopes had only increased. It wasn't just the United States that was producing them; the Atomic Energy Research Establishment (AERE) in Britain had launched their own export scheme of radioisotopes, produced in their Graphite Low Energy Experimental Pile (GLEEP), which went live in August 1947, just over a year after building had begun. Like at the Oak Ridge facility, these radioisotopes were highly prized by medical establishments and industry, and monthly shipments rose from around 100 in 1948 to 800 in 1951.[141]

Radioisotopes were used in two main ways – as tracers and as sources of radiation for treating diseases, especially cancer. The drive to find a cure for cancer was a dominant quest throughout the early-twentieth century.

As the 1920s and 30s progressed, radiotherapy evolved and became more powerful. However, the cost and rarity of radium salts remained a barrier to medical experimentation, and concerns around its safety, both for practitioners and patients, had been raised. This was especially important after the reports of the deaths of the group of industrial workers known as the 'Radium Girls' were circulated in the press and the medical journals. These women, who marked out watch faces with glow-in-the-dark radium paint in factories across North America, were estimated to have ingested between a few hundred to a few thousand micrograms of radium in a year during the course of their work.[142] They suffered terribly, and many died from radiation sickness, before a public outcry ended the practice of using saliva to form a point on the brushes used.

The first artificial radioisotope to be used clinically was radio sodium. First generated in 1934 in the Rad Lab at Berkeley, California, it was created by bombarding table salt with deuterons in the cyclotron. Ernest Lawrence claimed that the properties of this substance were not only 'superior to those of radium for the treatment of cancer', but available at a fraction of the cost.[143]

In 1948, *Popular Mechanics* reported that radioactive gold, made in the reactor pile at Oak Ridge, was being injected into leukaemia and cancer patients by Dr Paul F. Hahn, associate professor of biochemistry at the Vanderbilt University School of Medicine. While it didn't report on how successful it was, the article did say that the treatments eliminated 'many of the undesirable features of radium'.[144]

Unlike radium, which has a half-life of 1,600 years, the majority of radioisotopes had a short half-life, did not concentrate in the bones, and did not release the hugely damaging alpha particles.[145] Therefore it was reported that, 'synthetic radium' extended the treatment due to the 'fact that its radioactivity is lost within a few hours after its production, whereas that of radium persists indefinitely, presumably for thousands of years. Injection of radium into the human body is barred by this fact, since it acts as a poison. The effects of artificial radium would be terminated within a short time.'[146]

The second use of radioisotopes was first developed by Hungarian radio physicist George de Hevesy, who, in 1923, used an electroscope to observe how bean plants absorbed the naturally occurring isotope lead-212 solution through their roots, stems and leaves. Further experiments were conducted using animals as testing subjects to see how bismuth-210 was circulated through the body.

This technique is where a small amount of a radioisotope is introduced into a system or substance of interest, such as a patient's body. As the radioisotope decays, it emits radiation that can be detected and monitored. By monitoring the radiation emitted, researchers can track its movement through the system and observe how it interacts with other substances or processes. This allows them to study biological processes, diagnose medical conditions, monitor industrial processes, and more.

Joseph Hamilton expanded research into radiosodium by feeding small amounts of it to healthy subjects and monitoring the absorption rate. To do this, the subjects would drink a radioactive salt solution and place one hand around a Geiger–Müller counter encased in a lead cylinder. Hamilton discovered that radiosodium could be detected within two-and-a-half to ten minutes in the shielded hand after ingestion, and absorption appeared to be complete after as little as three hours in some subjects, or ten in others. Alternatively, radioactive sodium could be injected into the vein of an arm.

While it was hoped to use radioactive sodium as a curative, it became clear that its real benefit was for scrutinising the vascular system and detecting the location of blocked arteries.[147] The importance of this was not missed by the AEC, who in their 1948 semi-annual report stated: 'As tracers, they are proving themselves the most useful new research tool since the invention of the microscope in the 17th Century: in fact, they represent that rarest of all scientific advances, a new mode of perception.'[148]

An article in Life magazine called 'Atomic Progress' detailed various contemporary uses, including injecting radiocarbon into a cow – dubbed a 'Hot Cow' due to the measurement of the radioactive carbon dioxide it exhaled. Another application was 'Hot Fertilizer', involving radioactive phosphorous to assess the effectiveness of different fertilisers on the market. Other uses were in tracing the migration of mosquitoes and the habits of rodents, as well as testing floor waxes, tyres and the lubricating quality of oil.[149] The technique of feeding hens a 'radioactive mash' was used to find out how eggs developed.[150]

Senator Clinton P. Anderson, who was Chair of the Joint Commission on Atomic Energy, also laid out a benefit of the tracer labelling technique and eggs in a speech in 1955:

> If you believe your wife isn't getting the breakfast dishes clean, test how well she does her work. You merely feed some radioactive phosphorus to chickens who lay radioactive eggs: your wife fries eggs; you eat them but leave a little egg on the dish; she washes the plate and you check [with a counter] to see if radioactive material remains ... That's the way a dishwasher manufacturer tested the efficiency of ... a new dishwasher ... Radioisotopes did the job by finding traces of dried fried egg on a plate – and if you want to test it in this modern way at home, I can supply new isotopes and you can guarantee you'll either get a new dishwasher or a new wife, but you'll be doing it on a scientific basis.[151]

It wasn't just dishwasher manufacturers who used this approach. Companies like Bendix Home Appliances advertised that their Bendix 'Tumble Action Automatic' washer had been rigorously put through its paces and that its 'ATOMIC TESTS used Geiger Counters to accurately measure amount of dirt removed from cloth impregnated with radioactive soil'.[152]

Absorbine Jr, 'America's Number One Athlete's Foot Relief', used this approach in their advertisement: 'Now modern radioactive isotope tests prove the fungus-destroying ingredients in Absorbine Jr, are absorbed right into the skin.'[153] Vicks VapoRub promised mothers the effectiveness of their products as proved by 'atom tracer tests' carried out in their 'laboratories of atomic medicine'.

One of the most remarkable adverts of this period was by the beauty company Dorothy Gray for their Salon Cold Cream. In an eight-page *Atomic Test* booklet and a television advert it was extolled in great detail how they were using 'peacetime atomic research to measure and compare the cleansing powder of soaps and creams':[154]

> As most of us know, a Geiger counter can 'detect' even the slightest amount of radioactive material. This gave the laboratory scientists an idea. Why couldn't the cleansing power of soap and creams be determined by measuring the amount of soil they left on the skin after cleansing? This is exactly what was done in these tests!
>
> A soil consisting of dirt and make-up similar to what's found on a woman's skin during the day, was made just radioactive enough so that its presence could be measured by a Geiger counter. This soil was applied to the tester's skin.[155]

Using sterile cotton-tipped swabs, the scientists who conducted this experiment on behalf of Dorothy Gray applied the radioactive soil to the tester's skin. They then covered each spot with different cleansing products. After applying soaps and creams, they wiped them away with tissue, removing as much dirt as possible:

> Now came the critical moment ... the most dramatic and revealing step in the tests. The time has come for each soap and cream to prove how good it was! The Geiger counter was now used to determine how much soil was actually left on the skin after it had been cleansed with each of the soaps and creams.
>
> The laboratory scientist pressed the Geiger counter detector-tube against the tester's cheek where the skin had been cleansed. As the tube 'detected' the soil remaining after cleansing, the Geiger counter 'clicked' and its needle recorded the findings on the dial!

Dorothy Gray Salon Cold Cream cleansed skin up to 2 ½ times more thoroughly than anything else tested! Here was clear, scientific proof, that Dorothy Gray Salon Cold Cream definitely out-cleans ordinary cleansing methods. It left skin more completely, more beautifully clean![156]

While they didn't mention what they used to make the 'complexion soil', it was likely to have been carbon-14, a radioactive isotope of carbon, which was widely used in the cosmetic industry. The normal practice for experimentation was to use animal skin. In this instance Dorothy Gray chose to use a real person – although we can hope that the experiment was simulated rather than actually smearing radioactive material over the unnamed model's face. If you want to watch this rather shocking advertisement it is on YouTube.[157]

Carbon-14 was one of the most valuable of all radioisotopes, having already been put to use by archaeologists to determine the age of organic materials like fossils or other artifacts. It worked by measuring the amount of carbon-14, which is naturally produced in the atmosphere and is taken in by plants and animals during their lifetimes. When they die, the carbon-14 in their bodies begins to decay, and by measuring how much is left, scientists can calculate how long ago they died. This process was invented in the late 1940s by Willard Libby, a professor of chemistry who worked on the Manhattan Project and was a commissioner of the AEC during the 1950s.

Carbon-14 also played a significant role in the development of Automated Teller Machines (ATMs) in the late 1960s. According to legend, John Shepherd-Barron, a British inventor, came up with the idea one day after being unable to withdraw money from the bank due to their restrictive hours at the time. He used the concept of a chocolate vending machine as inspiration and came up with the idea of a machine that dispensed cash.

Shepherd-Barron pitched the idea to the Chief General Manager of Barclays bank, sealing the deal for an initial six ATMs, all of this taking place over a pink gin – don't you just love these details! The first De La Rue Automatic Cash System (DACS), also known as Barclaycash, was installed outside a branch of Barclays in north London. It was opened officially on 27 June 1967, by the popular star of the British TV show *On the Buses* and local resident Reg Varney.

The system required pre-approved customers to go into their branch and get special vouchers, each worth £10 and valid for six months. To use the automated system, the customer would sign the voucher and place it into a drawer of the machine. The machine would then test and detect the carbon-14 stripe on the voucher. If they matched, the cash was dispensed in £1 notes in another drawer. The potential hazards of using radioactivity were dismissed by Shepherd-Barron: 'you would have to eat 136,000 such cheques for it to have any effect on you.'[158] He also capped the amount you could withdraw at £10, which was said to be enough for a 'wild weekend'.

DACS was an instant success among those special customers. However, rival cash-dispenser systems quickly emerged, and it wasn't long before it was replaced by the reusable bank card and four-digit PIN we are now more familiar with.

ATOMIC CITIES

6

The successful Soviet testing of a plutonium atomic bomb code-named 'Fast Lightning' but dubbed 'Joe One' by the Americans, at Semipalatinsk, Kazakhstan, in August 1949, coupled with the escalation of the Korean Conflict in June 1950, as American troops were sent to aid South Korea, prompted an intensified focus on weapons development to beat the perceived threat of the spread of communism. As the United States reassessed its global superiority, it developed a multi-billion dollar expansion programme of production facilities at sites like Oak Ridge and Hanford. It also revived proposals for a continental test site and accelerated development of the highly destructive hydrogen bomb, which used a combination of fission and fusion (where two atoms merge to create a heavier one) to create an explosion.[1]

The US testing site in the Pacific had been successful at keeping the bombs well away from the American population, but it was difficult to conduct an operation at such a distance and the cost of transporting people and equipment was high. There was also a real concern that the site was not very secure. The AEC and the government knew they would have to work hard to convince the public of the importance of setting up a test site actually on the US mainland. President Truman wrote in his memoirs that he did not want to frighten people about 'shooting off bombs in their backyards'.[2]

A top-secret programme, 'Nutmeg', was launched to scout for potential locations, and after careful consideration, the Nevada desert emerged as the favourite choice.[3] The place chosen for the Nevada Test Site (NTS) was situated in the vicinity of a former air base, approximately 65 miles south-east of Las Vegas and 300 miles away from Los Angeles. This location met the criteria of being away from major population centres, and it was deemed highly secure.

But while the proposed area was fairly isolated, it was not uninhabited, with 100,000 people scattered in the rural area around the testing site.[4] It was also on Shoshone land, and the Shoshone people were forcibly displaced from their ancestral territory. Furthermore, the people living there would later bear the brunt of the radiation fallout from the tests.

In December 1950, President Truman gave his approval for the testing site, and the following year, Camp Desert Rock was built to accommodate the nearly 100,000 soldiers who would be stationed there over the course of the 1950s.[5] Being part of the test series earned them the moniker of 'Atomic Soldiers'.[6] Upon arrival at Camp Desert Rock, they were provided with an 'Information and Guide' booklet that emphasised the importance of strict security measures. Each page was stamped 'RESTRICTED' and contained numerous warnings against divulging classified information. It also emphasised: 'To assist in maintaining the security of Exercise Desert Rock it is desired that you maintain secrecy discipline regarding classified information observed here. Everyone will want to know what you have seen – officials, friends, and the enemy.'[7]

The guide also included warnings about health hazards, not related to radiation, but rather from the reptiles and poisonous insects in the area.

Casinos, bars and hotels in the area extended a warm welcome to the soldiers involved in atomic testing, with signs that read 'Welcome Atomic Soldiers' and offered discounted prices to those who arrived in town wearing their uniforms. The influx of visitors was welcome because Las Vegas was not quite the bustling place we know today. In the early 1950s it was still a relatively small city, with only around 25,000 residents and under a million tourists a year.[8]

The initial growth of Las Vegas from a quiet railroad town to a burgeoning tourist destination can be attributed to several factors, including the legalisation of gambling in 1931 and the introduction of the 'six-week cure' divorce. Nevada's divorce law required only a brief period of legal residence, the shortest time in the nation, which made it a popular destination for those seeking a quick split. Vegas wasn't just about ending marriages, of course, but was also a popular wedding destination, with, according to an article in *Business Week*, 2,944 divorces and 7,602 weddings in 1944 alone.[9]

The growth hadn't been without controversies, but by the early 1950s there were fifteen resort hotels, 38 commercial hotels and 286 motels. The four main hotels on the strip, El Rancho, Flamingo, Last Frontier and Desert Inn, would soon be joined by the Sahara, Sands and Stardust to name but a few as the town grew in popularity.

With the announcement of the establishment of the testing site there was, of course, concern about what this would do to the tourist industry, as well as the impact it would have on the residents' lives, with one Las Vegan quoted as saying: 'The Atomic bomb! No one asked us what we thought about it. Everyone is in uproar, for none of our officials know anything about this. All the gambling people are furious for, naturally, they fear that people will no longer come here.'[10]

On 27 January 1951, the first test of the Operation Ranger series was detonated at Frenchman Flat, a dried-up lakebed in the middle of the test site, and it was met with a sense of alarm.[11] The unknown loomed large, and the entire experience was unsettling, as reported in the New Yorker the following year:

> Most of the residents were awakened by tumbling window shades and shaking walls; some of them were tossed out of bed. Nobody was hurt, but one of the town's two daily newspapers, the Review-Journal, indulged its readers' dire expectations with the front-page headline 'VEGANS ATOMIZED'. All that day, there was worried speculation as to whether this might be only a tame curtain-raiser, but on the following morning a second shot came off and turned out to be no worse, and the atomized Vegans began to take their obstreperous neighbor in stride. Some of them expressed their relief by filing damage claims, of varying validity. Several homeowners declared that the shock waves had cracked the walls of their houses, but in more than one instance investigators found an accumulation of dust and cobwebs in the fissures.[12]

Despite initial concerns, the city of Las Vegas wholeheartedly embraced the atomic bomb tests. Newspapers reported that the residents of Las Vegas would wake their children to view the spectacle: 'We'd get up to watch it and hear it and watch the pink cloud go over. We thought it was something to see, something great.'[13]

As Doris Leighton, an administrative assistant at the Nevada Construction Company, told the New Yorker reporter Daniel Lang in 1952:

> Around five-thirty in the morning, the lights would start going on in my neighborhood. Some of us would come out on our

porches with cups of coffee and wait there. We'd be wearing
heavy wrappers, because the winter mornings were quite nippy,
you know. Sometimes husbands would back their cars out of
the garage and into the street to get a better view. They'd let
the motor run until the car was warm, and then their families
would come and join them. I used to see parents pinching
small children and playing games with them to keep them
awake. I guess they wanted to be sure their kids would see
history in the making. People all looked expectant but in different ways. Some, you could see, were afraid. They smiled and
acted nonchalant.[14]

The NTS also brought significant job opportunities to the area,
which greatly contributed to the expansion of Las Vegas.[15] The
operation and maintenance of the test site required a diverse
workforce, including scientists, clerical personnel, techni-
cians and other skilled workers, many of whom were based at
a purpose-built encampment – Camp Mercury – nearby. This
influx of employment opportunities attracted new people to the
area, stimulating economic growth and development.

The expansion of Las Vegas can be attributed in part to the
economic impact of the test site, which brought not only direct
employment opportunities, but also indirect benefits to the
local economy. The increased population and economic activity
associated with the tests contributed to the growth of businesses, services and infrastructure in the surrounding areas,
which helped transform it into a more diversified and vibrant
city beyond its reputation as a gambling destination.

And the tests also had a significant impact on the tourist
trade. Just two weeks after Operation Ranger began, the *Los
Angeles Times* published an article that indicated the local
businesses view: 'Even the Chamber of Commerce takes a
positive position. "These tests certainly haven't hurt Las

Vegas ... We're grateful Nevada is able to provide the facilities for these tests. We want to lend our support to national defence."'[16]

As the town and the tourists began to settle into the strange rhythm of the various testing series, atomic imagery gradually integrated into Vegas's culture.[17] Truly leaning into its new status, embracing all things atomic, this transformation was evident not only from signs on some of the casinos, including the unmistakable resemblance of the early Stardust and Flamingo signs to mushroom clouds, but also in the storefronts featuring slogans such as 'Atom Drops on All High Prices'.[18] The bomb became another part of the experience of visiting Las Vegas, merging with the town's glamourous reputation, the thrill of gambling and a pervading sense of otherworldliness. For businesses, the tests presented yet another avenue to market the region as a resort destination.[19]

To ensure that tourists were aware of the upcoming atomic tests, calendars were distributed by businesses confirming the dates and times. Road maps were also provided to highlight the best vantage points for watching the tests.[20] Some hotels even provided packed lunches for their customers to take to picnics at Angel's Peak, a mountain located 45 miles away.[21]

Many of the hotels, motels and casinos had atomic test offerings that were designed to bring in the public. At the Sands you could go to one of their 'Atom Watch' parties with breakfast served on the terrace.[22] And at the Atomic View Motel it was possible to lie by the pool and watch the tests from the comfort of your lounge chair.[23] The Desert Inn served up their version of the 'Atomic Cocktail' during parties in its third-floor lounge, the Sky Room. The tipple was made from equal parts vodka, brandy and champagne, with a dash of sherry.[24] Ted Mossman, who was the pianist in the panoramic glass lounge, remembered how the parties were:

Standing room only. They were drinking like fish. Some of them had cameras for photographing the flash – a thing they couldn't have done even if they'd been sober. It's too bright. Everyone wanted to sing ...

They sang as if they were on the *Queen Mary* and it was going down – loud, separate voices.

After a while, I couldn't take it anymore, so I improvised some boogie-woogie that I called 'The Atomic Bomb Bounce'. I kept playing it and playing it, until I thought my fingers would fall off.

Seven o'clock in the morning, we get word there's been a circuit failure out at the proving ground and the bomb's called off. They all started betting when the next bomb would be exploded – the week, the day of the week, the hour of the day.[25]

In the Venus Room at the New Frontier Hotel and Casino, there was a unique performer billed as 'the nation's only atomic-powered singer', none other than Elvis Presley, in his first Las Vegas appearance. *Variety* hailed him as an 'atomic-age phenomenon', while *Time* declared him to be 'hotter than a radioactive yam'.[26]

The publicity machine in Las Vegas proved to be remarkably successful in reshaping public perception of atomic testing, shifting it away from its initial associations with death, destruction, and towards a more glamorous connotation. The concept of the atomic bomb being 'hot' or 'radiating' took on a new connotation beyond its literal sense, becoming associated with modernity, excitement – and even sexiness.

This shift in perception was reflected in the emergence of atomic beauty competitions. While these competitions were not strictly based on traditional beauty queen pageants, with titles mostly being bestowed upon the women rather than earned through a competition, they still received a significant amount of publicity. Photographs of these queens, dressed in atomic-themed outfits such as swimsuits adorned with mushroom-cloud motifs, carrying Geiger counters or wearing atomic headdresses, were widely circulated through press releases and newspapers.

One notable example was the 'Miss Atomic Blast' competition in 1952, which saw Candyce King crowned the winner. She was described as radiating 'loveliness instead of deadly atomic particles'. Another winner was Lee Merlin, who was crowned 'Miss Atomic Bomb'. The photograph of her, which was taken by Don English, publicity photographer for the Las Vegas News Bureau, shows a smiling, beautiful woman wearing a cotton mushroom cloud attached to the front of her bathing suit, her arms raised into the air against the desert skyline.

But while the new atomic culture was being embraced, there was, of course, no way of hiding the physical effects of the tests, as residents made it clear in numerous articles: 'We would wake up in the mornings sometimes and the water would be sloshing out of the pool, just exactly the same thing as an earthquake. Chandeliers would swing, water would slosh out of swimming pools, the elevators inside the elevator transoms would swing back and forth.'[27]

Wilbur Clark, the owner of the mob-funded Desert Inn and as such someone presumably with nerves of steel, recalled after a particularly large shock that saw 200 damage claims sent to the AEC: 'That was another time we had an especially good take on gambling. Same for liquor. Hell, I took an extra drink myself.'[28]

Windows were a frequent casualty in both Las Vegas and Los Angeles, which despite being hundreds of miles away still felt the impact. The AEC settled city dwellers' claims promptly, which largely stopped public upset. Although one business haberdashers, Allen and Hanson, took a more direct approach. Following a test which saw the panes of their display windows shattered, they simply swept up the fragments, put them in a barrel outside the shop and posted a sign: 'ATOM BOMB SOUVENIRS – FREE!' They were soon snapped up by bomb tourists eager for a memento of their stay.[29]

Unbelievably, there were accounts indicating that the only perceived danger was from the sheer excitement of witnessing the tests, particularly for those driving in the vicinity, with one reporter reassuringly writing that: 'there is virtually no danger from radioactive fallout' but cautioned against automobile accidents because 'in the excitement of the moment people get careless in their driving'.[30]

The apparent safety of living near atomic tests was emphasised in many publications, both nationally and locally, with newspapers providing a number of reassuring headlines as the test series progressed: 'Use of Taller Towers ... Introduces an Added Angle of Safety', 'A-Bomb Incident "Proves" Safety', 'Radiation Danger in Vegas Nil'.[31]

Well-known reporters with national platforms regularly travelled to 'News Nob', which was ten miles from 'Ground Zero', to witness the tests.[32] Their reports were largely carbon copies of AEC talking points around safety, especially the absence of any radiation hazards.[33] And, indeed, the AEC and press sought to minimise fears and often downright ignored concerns from people living in the surrounding areas, later known as the downwinders.

While testing atomic bombs and their psychological effects on the atomic soldiers stationed there were important considerations, it was also deemed essential to prepare for what was considered to be an inevitable attack against the US. To this end, the AEC oversaw the construction of two replica neighbourhoods between 1952 and 1955 on the Nevada Test Site. These were called 'Doom Town' and 'Survival Town' respectively. As part of a series of tests known as Operation Teapot, the purpose was to simulate how a typical American community would fare during an atomic attack. To make this as realistic as possible they were built with fully furnished houses of different sizes, bomb shelters and electrical substations. Mannequins fully dressed right down to their underwear with clothes from the local J.C. Penney store were positioned in the houses, and additional mannequins were placed at varying distances from the epicentre of the explosion to assess their chances of survival.[34] Despite the different destruction rates of the buildings, the fate of the mannequins was pretty consistently horrifying:

> People played by dummies lay dead and dying in basements, living rooms, kitchens, bedrooms ... A mannequin mother died horribly in her one-story house of precast concrete slabs. Portions of her plaster and paint body were found in three different areas. A mannequin tot ... was blown out of bed and showered with needle-sharp glass fragments ... A simulated mother was blown to bits in the act of feeding her infant baby food.[35]

The cities of Las Vegas and Los Angeles also conducted drills for a possible attack, and it was soon determined that civil defence should be a national programme. To oversee this President Truman looked towards the National Security Resource Board. Later, these responsibilities were assumed by the newly formed Federal Civil Defense Administration (FCDA), which was created under an act signed into law by Truman on 12 January 1951.

Since it was presumed that major cities would be prime targets, the government implemented various programmes to educate citizens about the risks of such an attack and how to survive it. The main message was that survival was possible – if you knew the right precautions to take.

Along with regularly scheduled air-raid drills, there were books like *Pattern for Survival, You Can Beat the A-Bomb* and *Survival Under Atomic Attack*, which was published by the FCDA and distributed to 20 million Americans during 1950 and 1951.

The publications explained simple precautions that could help limit exposure to radioactivity in the event of an atomic attack, as well as ways to protect yourself during impact. Dozens of instructional films, pamphlets and posters were circulated throughout the United States.[36]

In 1951, the FCDA produced the pamphlet *Duck and Cover*. Along with the 20 million of these distributed, *Duck and Cover* was also made into an animation, an album and a radio programme. The film was first shown as part of the FCDA's 'Alert America' travelling show and was then shown in schools across the US.

The nine-minute-long film opens with an animated sequence featuring the character Bert the Turtle, who demonstrates how to protect himself by ducking into his shell when he sees a monkey wielding a lit stick of dynamite. The accompanying theme tune is sung by a chorus, encouraging viewers to follow the turtle's lead in the event of danger. At the first warning of attack, 'Like Bert, you duck to avoid the things flying through the air ... and cover to keep from getting cut or even badly burned.'

After the animated sequence, live footage is used to illustrate survival tactics in the event of an atomic bomb. The film advises children to duck down low, crawl under desks, and cover their necks with clasped hands to reduce the risk of injury. The film also suggests facing a wall that might provide protection during the attack.

The film ends with Bert summarising the safety instructions and asking a group of unseen children what they should do in the event of an atomic bomb. The children respond with the correct answer, reinforcing the message of the film.

During the early days of the Cold War the advice in *Duck and Cover* was pretty good – seeking refuge under a desk could indeed shield people from shattered glass and other debris. However, as the arms race intensified and both sides amassed weapons, including massive thermonuclear devices, the 'duck and cover' approach as a protective measure dwindled in usefulness. This new type of bomb was at least 150 times more powerful than the one that destroyed Hiroshima. In 1959, the Office of Civil Defense declared the technique obsolete.

Although the government provided civil-defence training, other efforts largely fell short.[37] Some fallout shelters were designated in cellars of schools, libraries and suburban malls, marked by yellow-and-black metal signs on exterior walls.[38] However, these buildings could only provide protection for a small percentage of the population and were often inadequately stocked with food and supplies.[39]

The 15 September 1961 issue of *Life* magazine began with a letter from President John F. Kennedy to 'My Fellow Americans', summarising a speech he had previously given over national radio and television a few months earlier. Acknowledging how close the world was to another war he urged them:

> to read and consider seriously the contents of this issue of LIFE. The security of our country and the peace of the world are the objectives of our policy. But in these dangerous days when both these objectives are threatened we must prepare for all eventualities. The ability to survive coupled with the will to do so therefore are essential to our country.[40]

There were lots of adverts and guidance articles in the national press about how to construct your own private shelters. The magazine had a twelve-page spread instructing the public how to build four different types – in the cellar, in a concrete pipe in your backyard, a double-walled above-ground bunker, or a prefabricated shelter.[41] It was a message that was taken to heart by many and provided an opportunity for entrepreneurs.

In July 1959, Bomb Shelters Inc, whose business card featured a mushroom cloud and the tagline 'It will save your life', sponsored a radio contest. The selected winners were Melvin and Maria Mininson, who made headlines when they spent their fourteen-day honeymoon in a twelve-feet deep, six-by-fourteen-feet wide fallout shelter.

The Mininsons faced some technical difficulties during their stay in the shelter. They sweltered in 90-degree heat (32 degrees Celsius), as the concrete in the hastily constructed shelter had not cooled properly. In addition, the couple's can opener broke, forcing them to improvise with a pair of scissors and resulting in Melvin cutting his hand. Despite these challenges, however, the couple emerged from the shelter unscathed and still happily married.[42]

The importance of stocking your shelter with the right survival gadgets, as well as some creature comforts – from long-lasting food to portable radios and blankets – was often emphasised. There was even the 'Fallout Shelter Window', as advertised by the Nashville Sash & Door Company in 1962: 'It is now possible to add daylight to your Fallout Shelter and still maintain protection from radiation with this new FALLOUT SHELTER WINDOW. Although not a view window the special high light transmission glass of this window will provide an abundance of daylight within your shelter.'[43]

The building and stocking of shelters became a genuine consumer fad and cultural touchpoint. As *Time* magazine put it: 'At cocktail parties and PTA meetings and family dinners, on buses and commuter trains and around office watercoolers, talk turns to shelters.'[44] However, while there was endless coverage in newspapers and magazines and lots of public discussion, evidence indicated that few people actually prepared in any meaningful way against the possibility of atomic attack, and only around a million families in the United States actually built one.[45]

But while a personal shelter was a huge expense, there were a number of more affordable gadgets available that could help with any fallout – should you survive the blast. Geiger counters were of crucial importance, to test radiation levels after an attack. It was reported that they 'were becoming as much a household necessity as a toothbrush'.[46] As the singer Bob Dylan put it in his memoirs: 'The general opinion was, in case of nuclear attack all you really needed was a surplus Geiger counter. It might become your most prized possession, would tell you what's safe to eat and what's dangerous. Geiger counters were easy to get. In fact, I even had one in my New York apartment.'[47]

While a Geiger counter was a useful monitoring device, Flobar advertised their hand soap as being an essential piece of kit in washing away contamination: 'The Atomic Energy Commission recommends the use of detergents as beneficial because they promote effective and rapid washing away of radioactive particles that may settle on the body or become trapped in the skin oils.' Flobar, which was actually a refillable container made from Lucite, was recommended as necessary to 'BE PREPARED ... Protect your family and yourself by keeping FLOBAR DETERGENT on hand ALWAYS!'[48]

In a mere twelve years, the United States conducted approximately 193 tests involving atomic and hydrogen weapons. These tests took place in the Pacific region as well as the Nevada Test Site. Meanwhile, the Soviet Union carried out 86 tests on its proving grounds in Semipalatinsk, and on Novaya Zemlya, a collection of islands in the Arctic Ocean.[49] Britain conducted tests in the Montebello Islands, and then at Emu Field and Maralinga in Australia. And later, Operation Grapple was a series of British tests at Malden Island and Kiritimati in the Pacific.

Initially there was strong evidence that the American public supported the testing programme, with results from surveys, such as nationwide Gallup polls, and electoral data clearly indicating that overall, the public wanted continued testing of weapons and advocated developing bigger ones, including the H-bomb.[50] Those that didn't could find themselves accused of being not only anti-American, but part of a communist plot. As the decade progressed, this sentiment began to change. In the period which historians Scott C. Zeman and Michael A. Amundson refer to as High Atomic Culture (1949–63), we see the collapse of the 'Cold War Consensus', which was characterised by general support of how the government handled the Cold War, and a move towards Americans becoming increasingly disparaging of atomic weapons.[51]

This growing sense of unease was sparked by incidents like the effects of Castle Bravo, a thermonuclear weapon detonated on 1 March 1954 at Bikini Atoll. It was the equivalent of a thousand Hiroshima-sized bombs and was the largest weapon the US had ever tested.[52] However, Bravo was not only unprecedented in size, but also in danger. The immense release of radioactive debris carried by unpredictable wind from the blast had far-reaching effects on the surrounding areas. A Japanese fishing boat, *Lucky Dragon No. 5*, over an hour-and-a-half away from the test site, was covered with ash from vapourised coral, which was referred to by the Japanese press as 'ashes of death'. The sailors on board became sick, with one tragically losing their life from acute radioactive sickness due to the exposure.[53]

The aftermath of the Bravo test was also catastrophic for the inhabitants of the surrounding areas, particularly the Rongelap and Utirik atolls, over a hundred miles away from the test site. The people of Rongelap suffered from a range of health issues including nausea, skin rashes, vomiting and other alarming symptoms. The fallout was so acute that by late March, Bikini, Rongerik, Utirik, Ujelang and Likiep atolls were deemed off-limits, and the inhabitants were forced to relocate again.[54]

Adding to the horrors, government researchers evacuated some of the islanders and enrolled them in a secret medical experiment known as Project 4.1.[55] This clandestine study aimed to further the understanding of the effects of radiation on human beings. Shockingly, the islanders were unwitting subjects in this unethical experiment, which resulted in deaths and miscarriages, and long-term health consequences.

The growing debates surrounding radioactive fallout shifted focus towards the hazards associated with specific fission products, notably strontium-90 and iodine-131.[56] The Joint Congressional Committee on Atomic Energy was in the final stages of conducting a series of hearings that garnered significant attention from the media. These hearings primarily centred around the threat posed by strontium-90. Of particular concern was the potential contamination of food and drink as it became understood that a small amount of fallout could find its way into the food chain, through diffusion into the soil – and into humans. The discussions and deliberations received extensive coverage, amplifying public awareness and apprehension.

It was the first time that concerns around fallout had been acknowledged at such a high level. In fact, the term hadn't even come into common usage until around 1952, replacing the clumsy terms 'dissemination of radioactive materials by air and water' or 'the evil cloud'.[57] The AEC had known that there were concentrated areas of radioactivity after an atomic bomb had been exploded. While publicly providing assurances that there was no need to be concerned, the AEC had begun tracking the path of fallout clouds and monitoring the air. Their priority was to ensure that the tests proceeded without hindrance. So, the

Uranium glassware: Dubarry talcum-powder flask with signature scent Golden Morn (*left*) and Lillicrap's Hone, a block designed to prolong the life of safety razor blades. (Sonee Photography)

Atom Bomb perfume – an earthy and potent fragrance. (Sonee Photography)

Uranium mining share certificates

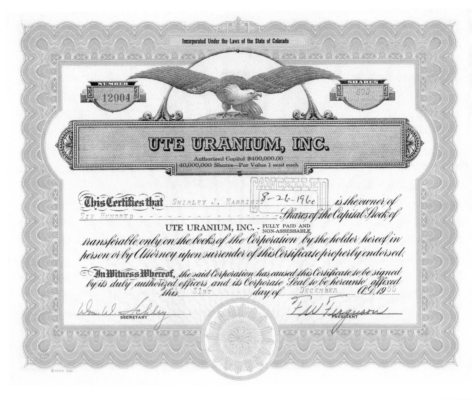

ACME URANIUM MINES, INC.

INCORPORATED UNDER THE LAWS OF THE STATE OF DELAWARE

NUMBER D12620

SHARES 1000

THIS CERTIFIES THAT *** ROBERT S. SPRAGUE *** is the owner of

*** ONE THOUSAND ***

full-paid and non-assessable Shares of the par value of $.01 each of the Common Stock of

ACME URANIUM MINES, INC.

transferable only on the books of the Corporation by the holder hereof in person or by duly authorized Attorney, upon surrender of this Certificate properly endorsed.

This Certificate is not valid unless countersigned by the Transfer Agent.

WITNESS the Seal of the Corporation and the signatures of its duly authorized Officers.

Dated: JAN 11 1957

Secretary

President

COUNTERSIGNED
COLORADO TRANSFER SERVICE, INC.
232 PATTERSON BUILDING
(DENVER, COLORADO)
TRANSFER AGENT

AUTHORIZED OFFICER

INCORPORATED UNDER THE LAWS OF UTAH

NUMBER 5578

SHARES 1,000

LIGHTNING URANIUM COMPANY

Capital Stock $125,000.00   12,500,000 Shares
Par Value 1c   NON-ASSESSABLE

Countersigned and Registered even date
Salt Lake City, Utah
By
Transfer Agent & Registrar

This Certifies that Hugh Coleman is the registered holder of ONE THOUSAND Shares

LIGHTNING URANIUM COMPANY

transferable only on the books of the Corporation by the holder hereof in person or by Attorney upon surrender of this Certificate properly endorsed.

In Witness Whereof, the said Corporation has caused this Certificate to be signed by its duly authorized officers and its Corporate Seal to be hereunto affixed this 8th day of July A.D. 1955

Secretary

President

RAY PRINTING CO., SALT LAKE

THIS CERTIFICATE NOT VALID UNLESS
COUNTERSIGNED BY TRANSFER AGENT.

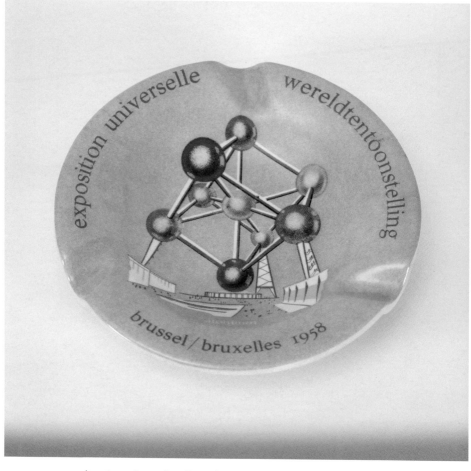

Atomium ash tray from Expo 58, an event that emphasised the peaceful
applications of nuclear energy. (Sonee Photography)

A group of anti-nuclear power badges – including the smiling sun design created by Danish activist Anne Lund. (Sonee Photography)

A group of pro-nuclear power badges. (Sonee Photography)

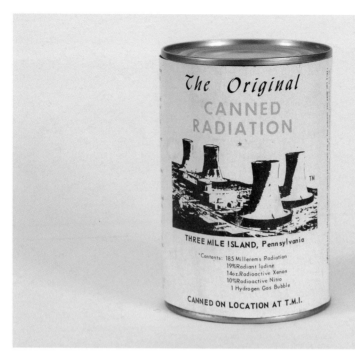

Three Mile Island Canned Radiation – canned air, to be more precise. (Sonee Photography)

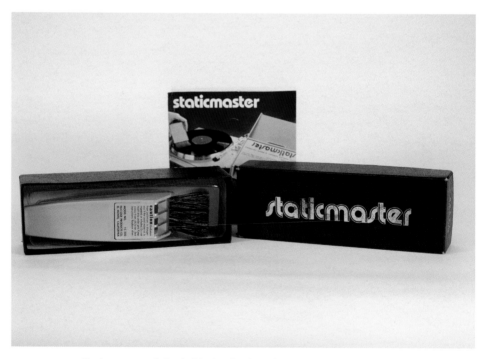

Staticmaster, made by the Nuclear Products Company of California, which contained radioactive polonium-210. (Sonee Photography)

Toxic Waste Nuclear Fusion sweets, Candy Dynamics, 2023. (Sonee Photography)

The Simpsons Interactive Nuclear Power Plant Environment with Radioactive
Homer, Playmates Toys, released 2000. (Sonee Photography)

Atomic tipples: Atomik Spirit from Chernobyl, and Nagurayama
'Yokikana' Sake from Fukushima. (Sonee Photography)

Author at the Grand Hotel Wien
in Vienna, the first home of the
International Atomic Energy Agency.

40,000 people living near the Trinity test were not informed or evacuated in 1945, and later, those living near or downwind of the Nevada Test Site were assured that their proximity would cause no 'detectable injury to health'.

Unease had erupted into a crisis in 1953 when a series of eleven nuclear test shots, known as Operation Upshot–Knothole, resulted in 252 kilotons of radioactive fallout passing over nearby roads and towns. Of particular concern was the effects of shot Harry – which was later nicknamed 'Dirty Harry'. The fallout was carried by winds over the nearby town of St George, Utah, and its surroundings. While residents were warned to stay inside, and highway roadblocks stopped cars for compulsory decontamination, there was no protection for the community's livestock, and thousands of lambs and ewes died. The AEC's assurance that this widespread carnage had been caused by mal-nutrition and cold weather failed to satisfy anyone.[58]

Fears around radioactivity were further aggregated by the embrace of a flawed model for radiation exposure, which was heavily influenced by a proponent of eugenics, geneticist Hermann Muller. Muller's research focused on the genetic mutations induced in fruit flies through ionising radiation expo-sure, which he had initially presented at the Fifth International Genetics Congress held in Berlin in the autumn of 1927.[59] He had found that even the smallest doses of X-ray radiation could cause genetic mutations, a process known as 'mutagenesis'. In 1946 he won the Nobel Prize in Physiology or Medicine for this research. Despite his work being characterised by limited data, a lack of consistent findings and the questionable extrap-olation from flies to humans, Muller still felt confident enough to conclude that mutation frequency was 'exactly proportional to the energy of the dose absorbed. There is, then, no trace of a critical or threshold dosage beneath which the treatment is too dilute to work.'[60] This model posits that even the smallest increment of radiation dose poses an elevated risk of cancer for humans.

Although Muller's conclusions were not universally embraced and other scientists charged that he had suppressed dissenting findings, his research laid the groundwork for the Linear No-Threshold (LNT) model of radiation exposure. In 1956 the US National Academy of Sciences and the British Medical Research Council (BMRC) recommended that the LNT model should replace the previous tolerance dose-response framework, which had been the guiding protection principle for the previous few decades and argued that there was a safe threshold of radiation exposure beneath which no untoward effects would occur.[61]

At the same time there was a shift in newspaper coverage, with even the most previously steadfast supporters of testing beginning to express their doubts.

The *Las Vegas Sun* wrote of a 'group of Nevadans' who lived and worked near the test site but were now 'confused and afraid' by something 'unseen' that 'makes their Geiger counter go crazy. Their eyes burn and sometimes the air has a chemical taste'. The piece concluded that they would like to know what was causing all the problems 'but nobody tells them anything'.

Playboy magazine expounded on 'a conspiracy of silence' around strontium-90. That this was a rather meaty topic for the magazine was addressed: 'This, you may now be saying to yourself, is an odd message to be appearing in a magazine dedicated, as PLAYBOY is, to life's good things, to the joy and fun to be found in the world: but these good things, this joy and fun, will cease to exist if life itself ceases to exist. And that is precisely what may happen.'[62]

There were films that captured growing fears, including the giant ants in *Them!* and the prehistoric monster *Godzilla*, which postulated what would happen if creatures or people were irradiated, either through fallout from atomic tests or, in the case of the Blake Edwards comedy *The Atomic Kid,* if you happened to be caught in one of the doom towns while a test was going off. Here the lead character, played by Mickey Rooney, is a uranium prospector who gets irradiated and gains special powers. While being treated in hospital he falls in love with his nurse. As they kiss, Rooney's wristwatch Geiger counter, which has been given to him by the scientists monitoring his condition, goes from 'danger' level to 'explosion', and his whole face glows brightly. There is also a great bit where his love interest, the gorgeous actress Elaine Devry, proclaims: 'I always pictured my dream man as being tall, dark and handsome. And then you come along – short, red headed and radioactive.' While Rooney is eventually cured, there isn't a happy ending for *The Amazing Colossal Man,* who, after being exposed to a plutonium explosion at the NTS, grows eight feet per day and terrorises the city of Las Vegas before being killed by the military.

This period saw a re-evaluation of the amount of non-essential radiation that people were willingly exposing themselves to on a day-to-day basis. Once-optimistic headlines now gave way to those that criticised what was termed a modern-day obsession with radioactivity, asking questions like 'Are Nuclear Tests Safe?' and 'Are We Getting Too Much X-Ray?':

Most people regard with terror the threat of radiation from an atom bomb. Yet those same people casually expose themselves to radiations just as dangerous. They have shoes fitted with X-ray machines, submit minor skin blemishes to X-ray treatments, have X-ray pictures made of various organs in the body.

Unaware of danger, a well-intentioned mother may drag her child from shoe store to shoe store, exposing him at each stop to radiation which would alarm any physician. To top matters off, that same day, the child may have X-ray pictures of his teeth made, plus a radium treatment for some throat ailment. We live in a radiation age, and we'd better learn its hazards.[63]

One of the technologies that came under fire was the fluoroscope, a machine that utilised X-ray technology and could be found in many shoe shops during the 1940s and 50s. This had been invented by Jacob J. Lowe during the First World War and had proved to be an invaluable tool for assessing the feet of military personnel without the need to remove their boots. This allowed for faster processing of large numbers of individuals.[64]

After the war, the device was adapted for shoe-fitting purposes, taking advantage of the surplus of portable X-ray units after they were no longer needed by the military. Consequently, the shoe-fitting fluoroscope became a common fixture in shoe stores in various countries, such as the United States, United Kingdom, South Africa, Canada, Germany and Switzerland, under different brand names.[65] By the early 1950s, the United States had an estimated 10,000 operating units, while the United Kingdom and Canada had approximately 3,000 and 1,000 units respectively.[66]

These wooden box-shaped devices featured a hole where children would place their feet. To view the X-ray images of the feet and shoes, one would look down through a viewing porthole situated on top of the fluoroscope. Additionally, there were two other viewing holes on each side for the sales assistant and accompanying adult.

However, concerns regarding the radiation exposure rates arose. A study conducted by Dr Charles R. Williams from the Harvard School of Public Health in Boston revealed that some machines exposed children's feet to nineteen times the amount of radiation per second that the US Bureau of Standards considered safe for a week's worth of exposure for actual scientific

X-ray workers. Operators were also subject to a huge amount of radioactive exposure, especially those that reported overseeing as many as 400 shoe fittings on busy days.[67]

By the early 1950s, several professional organisations, including the American College of Surgeons, New York Academy of Medicine, and the American College of Radiology, issued warnings against their continued use.

It took many years for their use to be discontinued but by 1970, shoe fitting X-ray units had been banned in 33 US states, and strict regulations in the remaining seventeen states made their operation impractical.[68]

There were also the reports of a search conducted as part of the international study 'Radium in Humans', which were published in newspapers in 1959. This study, which was sponsored by the AEC and included teams from the Massachusetts Institute of Technology and the Argonne National Laboratory, looked to find people who had drunk radon-infused waters, taken radium tablets or used uranium-ore treatments. While the aim of the study was to establish the long-term impacts of exposure to radioactivity, the rather sensationalised reporting served only to raise public concern. And, because of the popularity of radioactive consumer products in the recent past, there were plenty of potential dangers lurking in people's cupboards.

So, there were concerns around Fiesta Tableware, which had been reintroduced by Homer Laughlin in 1959. The only real difference between the new version and the old was that instead of using natural uranium, it now contained depleted uranium, a by-product of the uranium enrichment process. It was a popular revival, as people genuinely loved the range, and in fact still do – it is highly collectable! But the 1960s saw dramatic headlines like 'Hot Pottery Breakup', 'Radioactive Dishes Buried in Backyard' and 'Radiation Danger in the China Closet'.

These consumer panics, mingled with concerns around fallout and continued testing, led to protests and the establishment of organisations determined to eradicate the scourge of atomic bombs. In June 1957, social activist and author Ammon Hennacy embarked on a twelve-day protest against atomic testing at the offices of the AEC in Las Vegas. This act of protest, one of the first, garnered attention from various national media outlets, amplifying the reach and impact of Hennacy's demonstration and encouraging others to take up the cause.[69]

Just six weeks later, a collective named the Committee for Non-Violent Action Against Nuclear Weapons organised a demonstration at the NTS. Over 30 peace activists, predominantly Quakers, voiced their opposition to the 'senseless folly' of atomic testing. On 6 August, the twelfth anniversary of the Hiroshima bombing, eleven of this group attempted to enter the test site at Camp Mercury. However, they were apprehended and arrested by the local police. Although a local judge found them guilty of trespassing, their sentences were suspended under the condition that they refrain from committing further crimes.[70]

In the spring of 1958, there were several attempts by activists to sail into the testing zone of the Marshall Islands as a form of protest. One notable incident involved a group of Quakers aboard a 30-foot ketch named the *Golden Rule*. Their intention was to sail from Hawaii to Eniwetok, aiming to protest against the scheduled tests later that year. Referred to in the press as 'atom lopers', their journey was abruptly halted when the US Coast Guard detained the sailboat just two miles off the coast of Honolulu.[71] Despite the fact that they were not successful in their original mission, they secured a lot of press attention.

While there were small-scale 'Ban the Bomb' protests and resistance through non-participation in civil-defence drills, one of the highest-profile protest groups of this period was founded in 1957.[72] The initiative was spearheaded by Norman Cousins, editor of the *Saturday Review*, and Clarence Pickett, secretary emeritus of the American Friends Service Committee.[73] The first meeting was held at the Overseas Press Club, in New York, and was attended by people representing various fields such as business, science, labour, literature and the church. The group decided to call itself the National Committee for a SANE Nuclear Policy.

SANE sought an immediate cessation of nuclear testing and an international ban. They raised awareness through publications like *The Effects of Nuclear War*. They placed advertisements in newspapers that warned 'We Are Facing a Danger Unlike Any Danger That Has Ever Existed'.[74]

The response was encouraging, with widespread support gathering momentum within weeks. By the end of its first year, SANE boasted approximately 130 chapters and 25,000 members, solidifying its position as a prominent voice in the anti-bomb movement,[75] and swiftly growing to become the largest and most influential peace group in the United States.

In April 1958, SANE engaged in its first significant action. Over the course of several days, participants gathered in New York City for a demonstration. Some of the protesters had walked from as far afield as New Haven, Philadelphia and Westbury, Long Island, converging at UN Plaza as a symbolic gathering point. This event marked an important milestone in SANE's activism, but there were other groups involved in the march, including the Industrial Workers of the World, the War Resisters League, and the Women's International League for Peace and Freedom.

SANE also orchestrated the Madison Square Garden Rally on 19 May 1960, which drew an impressive attendance of 17,000 people, including notable guests and speakers such as Eleanor Roosevelt and Harry Belafonte, who added their influential voices to the cause. Following the rally, a powerful demonstration unfolded as 5,000 people marched through Times Square, making their way to the United Nations building.

One of the most instantly recognisable examples of these protest movements was the Campaign for Nuclear Disarmament (CND), formed in 1957 following a fire that had occurred at one of the Windscale Piles in the north-west of England. The fire in Unit 1 burned for three days and released radioactivity, which spread across the UK and the rest of Europe. One of CND's first activities was organising the Aldermaston March, which took place in April 1958. Starting in Trafalgar Square, London, and covering a distance of 50 miles, the protest finished three days later, at the Atomic Weapons Research Establishment (AWRE) in Aldermaston, Berkshire, a massive facility dedicated to nuclear research and the assembly of atomic and hydrogen bombs.[76]

The iconic CND symbol, which was designed by Gerald Holtom in 1958, incorporates the traditional semaphore alphabet's letters N (uclear) and D (isarmament). The initial badges were crafted from white clay with the symbol painted in black and were distributed with a note explaining that in the event of atomic war, these fired pottery badges would be among the few human artefacts to survive the inferno.[77]

While CND was primarily known for its symbol, the group also produced different badges conveying various messages. These emblems became a creative way to express opposition under CND's collective umbrella. Examples include 'Cat Lovers Against the Bomb', 'Dog Lovers Against the Bomb' (with the catchy slogan 'Dogs Not Bombs!'), 'Vegetarians Against the Bomb' and 'Taxidermists Say "Stuff the Bomb"'. These diverse badges allowed people with different interests and backgrounds to join the movement and make their voices heard, while also fostering a sense of unity. Badges with slogans such as 'Better Active Today than Radioactive Tomorrow' or 'Nuclear Power

Affects Your Sex Life' went further, using catchy phrases and playing on people's fears with arguably false claims that encouraged radiophobic emotions.

The marches and protests became powerful symbols of the collective desire for nuclear disarmament, demonstrating the widespread support among the public. And they soon spilled into more peaceful uses of atomic energy – an industry that was only getting started.

NUCLEONICS 7

The potential of uranium piles to provide energy had not been limited to the pages of the popular press, but there was a lot of research and developmental work needed to turn fantasy into a reality. This wasn't just progress for progress's sake – the Second World War had seen real shortages in everything from gasoline, which had been rationed, to the availability of easily mined coal. The consideration of the increased importance of energy production in post-war policies had started even before the war had ended, with the *Piles of the Future Review* and the *Prospectus on Nucleonics* both written in 1944.[1]

Neither of these reports had been particularly confident about the possibility of an energy revolution powered totally by uranium, with the *Prospectus on Nucleonics* concluding:

Even though the possibility of obtaining an unlimited quantity of nuclear energy, and making coal, oil, or falling water obsolete as sources of energy, lies in an as yet dimly perceived – but by no means fantastic or necessarily remote – future, there is no uncertainty whatsoever about the tremendous superiority of nuclear power over all other sources of energy as far as the concentration of energy and the possible intensity of its release are concerned.[2]

This assumption was largely based on an inaccurate estima-
tion of how much uranium ore there was in the world and the
technology available at the time for extracting uranium from
lowergrade ores.[3]

Despite the few coolheaded voices in those early, heady days,
the possibilities of atomic energy seemed endless. And with the
end of the war, science writers were free once again to speculate
about the practical applications, and they certainly did that!

Gerald Wendt in *The Atomic Age Opens* predicted: 'Generations
millenniums hence may look back upon these years when atomic
energy was first put to work in the same spirit in which we
now think of the less well documented occasion when man first
learned the use of fire.'[4]

In the book *Almighty Atom: The Real Story of Atomic Energy*,
author John O'Neill was so sure in the potentiality of uranium
energy that he predicted the extinction of the coal industry:
'With the coming of atomic energy, coal faces serious competition, so serious that its usefulness as a fuel and a source of heat
appears to be approaching an early end.'[5]

Some possibilities were rather creative – cartoonist Charles
Pearson added a touch of humour to the subject through a series
of drawings featured in the *Science Illustrated* article 'The Coming
Atomic Age Offers Some Awesome Possibilities' in February 1949.
With his knack for satire, Pearson illustrated potential pitfalls of the
atomic age. Among the depictions, 'Atomic Dry Cleaning' was portrayed as a potential hazard, humorously warning of 'spontaneous
decomposition'. Another drawing, captioned 'Atomic Baby Care',
comically depicted parents and baby encased in metal armour, with
a feeding flap for the infant. And another panel predicted that short
skirts would soon become fashionable again because the longer
ones were more likely to pick up atomic dust.[6]

Writers were not only fascinated by discussions on applications of atomic energy, but also by the concept of atomic

propulsion. The idea of vehicles, including large vessels like ocean liners, being powered by uranium had been proposed as early as the 1940s. However, it was after the conclusion of the Second World War and the deployment of atomic bombs that these ideas gained significant traction and captured the imagination of the public and writers alike.

Dr Langer followed up his 1940 'Fast New World' article with another one for *Popular Mechanics* – 'The Miracle of U-235' – the following year. Here, among predictions of 'universal comfort, practically free transportation, and unlimited supplies of materials' he also speculated that 'a power plant the size of a typewriter will be available', providing the ability to 'drive 5,000,000 miles without refuelling'.[7]

David Dietz, in his book *Atomic Energy in the Coming Era*, envisioned a future where atomic-powered airplanes would transport several thousand passengers, providing cabin space equivalent to a luxury liner.[8] However, even the writers themselves acknowledged that these ideas were largely fantasies. Even amid the excitement surrounding atomic propulsion, a sense of caution still prevailed. While the concept was intriguing, practical considerations and safety concerns tempered the enthusiasm and expectations of such endeavours. It was recognised that certain safety measures would be necessary to shield drivers and occupants from radiation exposure. And those who claimed to have achieved this seemingly impossible feat of technology were later exposed as frauds or revealed as dreamers.

In the former category was certainly the British scientist John Wilson who, in late 1945, made a bold assertion that he had the means of 'atomic propulsion' that purportedly had the capability to power various types of machinery, including railroad engines, ships and even aeroplanes.[9] He further claimed to have developed the first atomic-powered car, a modified Singer Nine Roadster. According to Wilson, this vehicle 'did 1000 miles at a cost of 8d. and travelled at three times the speed of an ordinary car'.[10] He explained how it worked for a British Sunday newspaper:

> It cannot go wrong ... The apparatus consists of a metal flask under the dashboard of the car and a steel cylinder under the hood. I fill the flask with liquid air. In the cylinder is suspended a fragment of uranium and four secret chemical elements. Introduction of the liquid air releases atomic energy which produces oxygen and hydrogen. These are fed as fuel directly into the cylinders, by-passing the carburetor. Combustion takes place in the ordinary way.[11]

To prove his claim a test was arranged in front of motoring correspondents, scientific writers and the Minister for Fuel and Power, Emanuel Shinwell. But on the day Wilson turned up without the car, reporting that it had been 'sabotaged' outside his office in Regent Street. Wilson didn't know who had done it but he 'suspected people apprehensive of the effect his invention would have upon industry'.[12] He later failed to make good on his promises to produce another one, and a court found that he was a fraud. He was sentenced to 21 months in prison for obtaining £250 by false pretences through selling shares in his invention.[13]

Despite the nonexistence of Wilson's nuclear-powered car, public interest in atomic automobiles remained high. An article in *Autocar* magazine predicted: 'There is now little doubt that the petrol-driven car as at present understood is entering upon the last years of its existence. The car of the not-so-very distant future will be driven by atomic power.'[14]

In the subsequent decade, several rather far-fetched attempts emerged for the development of these much-desired atomic-powered vehicles.

There was the Studebaker-Packard Astral and the Arbel-Symétric. The Arbel featured an innovative 'Genestatom' generator powered by used nuclear fuel. On the other hand, the Simca Fulgur, exhibited at the 1959 Geneva Auto Show, aimed to provide a glimpse into the future of automotive design in the next century. Beyond atomic propulsion, this concept car showcased a plastic bubble top and concealed wheels, adding to its futuristic appeal.[15]

The most infamous of all these concept cars was the Nucleon, developed by Ford in 1957. This car featured a unique power system, with a compact nuclear reactor situated in the rear. The nuclear reactor would harness the heat generated through fission to convert water into steam, and then propel turbines to provide power to the vehicle. Impressively, it was claimed the Ford Nucleon would have a range of 5,000 miles before requiring a charge, with spent fuel rods being replaceable at service stations. Although the Ford Nucleon never reached the production stage, several concept models were constructed to showcase its innovative design and potential.[16]

The history of the car was already one steeped in the evolution of the means of powering it. During the 1880s, automobiles had begun to make their presence known on the roads, albeit in small numbers. However, as the twentieth century approached, three distinct automobile types emerged. Gasoline-powered vehicles were relatively scarce compared to the more prevalent steam-powered and electric-powered counterparts. Remarkably, it is estimated that approximately a third of all vehicles on American roads in the first decade of the twentieth century were powered by electricity.

Electric cars were popular at the time largely due to the advantages they offered over gasoline and steam-powered vehicles. This was mainly around the ease of starting, with gasoline cars requiring manual cranking while steam power could be a time-consuming process, sometimes taking up to 45 minutes. Electric vehicles, on the other hand, started up quickly, were lighter to operate and did not involve complex gear-shifting mechanisms.

With the increasing popularity of electric cars, particularly among women, to whom they were heavily marketed, the need developed for convenient battery-charging methods. Electric cars of that era had limited range, often capable of traveling only around 50 miles on a single charge. Consequently, various approaches were developed to address this challenge, including a series of service stations and garages equipped to facilitate this.

One notable solution was dreamed up by the US-based Electric Vehicle Company, which introduced a rental system for their cars. Under this arrangement, users would return their vehicles to a designated garage each night. The company would then take care of charging and servicing the vehicles overnight, ensuring they were ready for use the following day.[17] So, four decades later, the concept of periodically swapping out reactor rods as part of the future of automobile technology did not seem entirely outlandish. The idea of changing your electric batteries was also modelled on even earlier technology. Specifically, it echoes the system of stables that existed during the era of horse-powered transportation. In those times, it was customary to exchange fatigued horses with fresh ones during a journey to maintain continuous mobility.

By the mid-1930s, electric cars had largely faded from prominence, while gasoline-powered vehicles had emerged victorious in the battle for automotive supremacy around 1912, via a combination of the development of the 'world's most affordable motor car' – the petrol-driven Ford Model T – the invention of the electric starter and cheaper gasoline.

In the late 1950s and early 1960s, several automobile companies, including General Motors, Chrysler and Chevrolet, revisited the idea of electric cars. However, despite their efforts, none of these electric vehicles managed to capture the public's imagination or gain widespread popularity. And so another alternative, atomic-power, emerged as a potential solution to address the limitations of electric.

If they had been a realistic proposition, atomic automobiles would have offered certain advantages over electric vehicles, such as overcoming issues related to limited driving range, lengthy charging times and bulky batteries. But making a nuclear reactor safe enough and small enough for use in a car proved impossible.

However, despite the acknowledged challenges, the possibility of realising atomic-powered vehicles and transportation continued to captivate the imagination of many, as typified by an advert for 'America's New Railroad' placed by the Santa Fe Railway Company in the mid-1950s. The illustration shows a boy dressed in an astronaut costume complete with a see-through domed space helmet asking the desk clerk: 'I want a ticket on the Atomic Super Chief!'[18] The text of the advert explains: 'Of course there is no "Atomic Super Chief!" ... yet. But, don't sell American ingenuity and progress short. At the rate things grow new for you on the Santa Fe, you only need to squint your mind's eye just a little to see tomorrow streaming toward you down the track.'

While railway companies were busy emphasising their modernity as they addressed challenges from cars and planes, it is that '... yet!' that embodied the spirit of endless possibilities, and the belief in the transformative power of technological advancements persisted despite the challenges and uncertainties. Similar to the early excitement surrounding the commercial and medical applications of radium earlier in the century, there was a shared hope that a future powered by uranium and atomic energy was within reach.

While radioisotopes had immediately found practical utilisation, atomic energy took rather longer to be embraced.

Cautioning against atomic-energy daydreams, policy makers and scientists largely distanced themselves from the predictions in the popular press and put the widespread commercial use of uranium to generate energy firmly in the 'not today but soon' box. President Truman said: 'Atomic energy may in future supplement the power that now comes from coal, oil and falling water but at present it cannot be produced on a basis to compete with them commercially. Before that comes there must be a long period of intensive research.'[19]

The AEC endorsed a cautious yet comprehensive plan for the advancement of nuclear energy. In line with this, a select few industry partners were granted access to evaluate the prospects of commercial development within the secure confines of the AEC's facilities.

During this period, a number of reactor types were tested, employing various moderating and cooling materials.[20] Crucially, experiments were conducted which made use of the availability of enriched rather than natural uranium.[21]

At the National Reactor Testing Station (NRTS) near Arco, Idaho, three new reactors were built in the late 1940s. Of the new designs it was a breeder reactor, designated EBR-I, that became the world's first nuclear energy plant producing small amounts of electricity – well, at least enough to light up its own building by the end of 1951. EBR-I went on to operate for twelve years before being shut down in 1963. It was made a National Historic Landmark in 1965 and is still available to visit today – either in person or on one of their fabulous virtual tours.

⟨≈⟩

A new step forward in both reactor technology and peaceful uses of atomic power was delivered in a momentous speech to the

General Assembly of the United Nations. On 8 December 1953 President Eisenhower unveiled a new public policy aimed at managing nuclear energy on an international basis. This speech, dubbed by the press as 'Atoms for Peace', was witnessed by 3,500 delegates representing 60 countries and was broadcasted live on both radio and television, amplifying its global reach and impact.[22]

For the first half of the speech Eisenhower reflected upon the past and emphasised the devastating consequences that atomic warfare posed, stressing the 'hideous damage' that could be inflicted.[23] The second part of the twenty-minute speech was markedly more hopeful and optimistic, focusing on the potential of harnessing atomic energy for the betterment of humanity. The President emphatically declared: 'this greatest of destructive forces can be developed into a great boon, for the benefit of all mankind. The United States knows that peaceful power from atomic energy is no dream of the future. That capability, already proved, is here – now – today.'[24]

To expedite this dream, he made a proposal of an 'atomic pool':

> The governments principally involved, to the extent permitted by elementary prudence, to begin now and to continue to make joint contributions from their stockpiles of normal uranium and fissionable materials to an International Atomic Energy Agency. We would expect that such an agency would be set up under the aegis of the United Nations ... The most important responsibility of this Atomic Energy Agency would be to devise methods whereby the fissionable material would be allocated to serve the peaceful pursuits of mankind. Experts would be mobilised to apply energy to the needs of agriculture, medicine ... to provide abundant electrical energy in

power-starved areas of the world ... The United States would
be more than willing – it would be proud to take up with others ... the development of plans whereby such peaceful use of
atomic energy would be expedited ... the Soviet Union must,
of course, be one.[25]

With a nuclear arsenal of 841 bombs, the US could now afford
to turn its attention to the more peaceful applications of uranium.[26] By promoting the exploration of peaceful uses of atomic
energy, the US sought to present a more balanced image and to
detract from the perception of being solely driven by military
ambitions. The Soviet Union, however, charged the Eisenhower
government with using the programme to sidestep a complete
ban on the use of atomic and hydrogen weapons and shifting
focus away from disarmament discussions.[27]

While there are still debates around how much this was a
genuine attempt to reduce the threat of nuclear war and how
much was propaganda, the speech was met with great enthusiasm, resonating not only with the American public but also
with people around the world. Extensive media coverage contributed to the widespread diffusion of the speech and sparked
further discussions on the potential of atomic energy for peaceful purposes.

Recognising the need to visually illustrate these seemingly
unlimited possibilities, promotional films were also swiftly
produced. These films delved into the various applications of
atomic energy, showcasing its potential in agriculture, medicine
and other peaceful endeavours. The production of these aimed
to engage and educate the public about the positive prospects
offered by atomic energy.

Some 200,000 copies of a pamphlet containing the speech,
translated into ten different languages, were distributed.[28] The
US Post Office released a postage stamp, valued at three cents,

featuring one of the key takeaways of Eisenhower's speech: 'Atoms for Peace: To Find the Way by Which the Inventiveness of Man Shall Be Consecrated to His Life.'

While these were successful pieces of public engagement, by far the most impressive ambassador for the programme was the world's first nuclear-powered merchant ship, the NS *Savannah*. This was a joint project between the AEC and the Maritime Administration of the US Department of Commerce. Officially named by First Lady Mamie Eisenhower in 1959, the *Savannah* was powered by a reactor safely housed in a 35-foot diameter containment vessel accessed through an airlock. Containing 682,000 uranium-oxide pellets, the reactor was powered up in 1961. Not only a demonstration of the peaceful atom, the *Savannah* was also a passenger ship, luxuriously appointed and fully fitted for comfort with 30 air-conditioned staterooms, a barber shop, swimming pool and dance floor.

As part of its promotional duties the ship took part in a 'Nuclear Week' series of educational events in New York City, which included two episodes of the hugely popular *The Tonight Show*. After travelling to 45 ports in 26 countries and welcoming over a million people on board, it was retired and defueled in 1971. It had travelled almost half a million miles powered by only 70 kilograms of uranium, but operating costs still outweighed income from passengers.[29]

While the atomic pool never actually materialised in the way initially envisaged, 30 countries, including Brazil, Portugal and Turkey, eventually signed up to the programme and 40,000 kilograms of enriched uranium was allocated. This valuable resource was predominantly distributed to universities worldwide, where it was utilised in small, inherently safe research reactors.[30] Atoms for Peace was not just about sharing fissile material; there was also technical assistance to help set up atomic energy programmes. An International School of Nuclear Science and

Engineering was set up in 1955, offering students seven-month courses in reactor and theory technology. The first session was attended by 40 students from twenty countries. By the time the school had closed in 1965, some 800 students had participated. The Atoms for Peace programme also led to the creation of the International Atomic Energy Agency, originally headquartered at the Grand Hotel Wien in Vienna.

The United States, however, faced a significant challenge due to the provisions of the Atomic Energy Act of 1946, which prevented the private industry involvement that was necessary to turn atomic vision into reality.[31] Recognising that a change was vital, President Eisenhower asked Congress to pass legislation 'making it possible for American atomic energy development, public and private, to play a full and effective part in leading mankind into a new era of progress and peace'. This call initiated comprehensive hearings and debates around the form this would take before a new Atomic Energy Act was eventually passed on 30 August 1954.[32]

This legislation paved the way for positive international cooperation and resulted in a new law that facilitated the exchange of nuclear technology with other nations. While defence and security remained important, the revised act placed greater emphasis on the development of peaceful uses of atomic energy.

Within the United States it allowed for the development of a commercial nuclear industry as a crucial national objective.[33] While industry did not acquire the right to possess fissionable material, the new Act redefined the atomic energy programme of the AEC. As well as ensuring the safety of commercial nuclear energy, it gave the Commission responsibility 'to encourage widespread participation in the development and utilisation of atomic energy for peaceful purposes'.[34] Now private industry, companies like General Electric and Westinghouse who were keen to develop new technologies, could build their own nuclear plants using fissionable material leased from the government, who still maintained control over ownership.[35]

The AEC was also tasked with preparing a report detailing their objectives for reactor development and the proposed timescale for achieving them. The report, submitted on 5 February the following year, outlined a comprehensive five-year plan that included research and development initiatives with an annual budget of $8.5 million. The eventual Power Demonstration Reactor Program (PDRP) invited private companies to design and build five different prototype reactors. The companies would own and operate them, but the AEC would provide funding, research and development on power reactors in its national laboratories, and wave fuel-use charges for the loan of fissionable materials for seven years.[36]

Initially, the response from industry in the US was cautious, primarily due to the complex and uncertain nature of developing this new technology. Despite the subsidies, the costs associated with this endeavour were high, and there was no guarantee of success. As the American Management Association symposium concluded in 1957: 'The atomic industry has not been – and is not likely to be for a decade – attractive as far as quick profits are concerned.'[37]

In a development that surprised many, the Soviet Union announced in June 1954 the initiation of Atomic Power Station 1 Obninsk, a water cooled graphite moderated reactor, the first civilian power station in the world.[38] APS-1, and the subsequent Soviet complex that included the building of 37,000 nuclear warheads and 30 nuclear power plants, utilised uranium oxide, much of which came from mines from conquered territories.[39]

Britain was also developing their atomic programme, which had begun with the construction of two nuclear piles in 1947, specifically to breed plutonium for nuclear research and weapons production. Work on the country's first commercial reactor began in 1953. The Calder Hall reactor was inaugurated by Queen Elizabeth II in 1956 and was claimed as the first nuclear power station in the world designed for large-scale electricity production for public use, although it also produced plutonium. Obninsk's status as being semi-experimental was determined to be the cause for its downgrade. In any case, Calder Hall surpassed the electricity output of the Russian reactor by a factor of ten and created enough electricity to power the needs of 15,000 people.[40] This new power source was greeted with great excitement by the press, with the *Daily Express* proclaiming: 'A-Power: It's Here! In Action! The First Dinners are Cooked!'[41]

The US's eventual first commercial nuclear energy plant was the Shippingport Atomic Power Station in Pennsylvania. Work began on 6 September 1954 after an elaborate ceremony, with President Eisenhower participating from his 'summer White House' in Denver, Colorado. In his speech the President said:

> ... for today ... we begin building our first atomic power plant of commercial size – a plant expected to produce electricity for 100,000 people. In thus advancing toward the economic production of electricity by atomic power, mankind comes closer to fulfilment of the ancient dream of a new and better earth ... through knowledge we are sure to gain from this new plant we begin today, I am confident that the atom will not be devoted exclusively to the destruction of man, but will be his mighty servant and tireless benefactor.[42]

After the address, which was broadcast to the guests and dignitaries at the Shippingport site via twenty television sets, Eisenhower passed an 'atomic wand' consisting of an enclosed radioactive source over a neutron counter. This flashed a signal to the site, over 1,200 miles away, and remotely activated a highlift, which scooped the first dirt.[43] The ceremony was also aired via radio and television to the public.

Lewis Strauss, Chair of the AEC, likewise addressed the crowd:

> Only a little more than a year ago it was believed that production of commercial amounts of electric energy from nuclear power would have to be demonstrated by the government ... and by the government alone ... before private industry would or could afford to take part in it. But so rapid have been the strides in scientific and engineering achievement that here, today the government ... that is to say, the people ... begin such an enterprise which is more fundamentally a pioneer adventure than the first railroad to penetrate the West or the first airline to span the continent.[44]

Shippingport was to be a pressurised water reactor (PWR), built by Westinghouse. In a PWR (which even in 2023 made up almost 70 per cent of all reactors in use globally), regular water is used both as a coolant and a moderator. The reactor core contains fuel rods filled with enriched uranium, which undergoes nuclear fission, releasing heat. The heat is transferred to the surrounding water, which remains under high pressure to prevent boiling. The pressurised water circulates through the reactor core and transfers its heat to a separate water loop, where steam is generated to drive a turbine and produce electricity, or, in the case of propulsion technology, turn the propellor shaft.

PWR technology was one of several types of reactors that had first been used for experiments for propulsion, under the US Navy's nuclear programme directed by Hyman G. Rickover.[45] While alternate reactor designs at the time included heavy water, helium gas, liquid sodium as a coolant and heavy water or graph-ite as a moderator, Rickover soon concluded that 'light water', which is ordinary H2O, was the best idea for the proposed nuclear submarines.[46]

Rickover, working with Westinghouse, designed a proto-type Submarine Thermal Reactor, and in January 1955 the USS *Nautilus*, the first nuclear-powered submarine, was launched. As the vessel launched Commander Eugene P. Wilkinson signalled the message: 'Underway on nuclear power.'[47]

It hadn't been right for cars and trains, but nuclear energy was perfect for submarines: the large amount of power allowed the vessel to run for long periods of time without needing to be refuelled.[48] As *National Geographic* magazine put in 1954, the 'range of operation will be limited not by fuel supply but by the crew's endurance, bottled oxygen, food and weapons.'[49]

It was the success of this project that generated the initial interest in nuclear energy for both Congress and private indus-try and played a significant role in the widespread adoption of PWRs in the US.[50] It was also the reason why Westinghouse, under the direction of Rickover, was chosen by the AEC to design Shippingport using the PWR technology.[51] It offered a workable model.

The construction of Shippingport was successfully com-pleted in 32 months at a cost of $75.5 million.[52] The station featured various essential systems, including four steam gen-erators heated by the reactor, a single turbine generator, and associated infrastructure such as used-fuel disposal facili-ties, laboratories, shops and administrative buildings. Unlike Obninsk and Calder Hall, which also had the dual purpose of making plutonium for the Cold War stockpile of nuclear weap-ons, Shippingport had no military mission and instead was largely developed to convince the public that nuclear energy was clean and safe.[53]

On 2 December 1957, which was only fifteen years to the day that CP-1 first underwent nuclear fission, the reactors at Shippingport underwent testing, a process known as going cold critical, to ensure their safe and efficient operation, and were then brought to full power 21 days later and connected to the grid the following year.[54] It was an important development, as Mark W. Cresap Jr., the President of Westinghouse Electric Corporation explained:

> The significance of Shippingport is twofold. It is a revolutionary type of full-scale electric power generation. It is also a full-scale training and testing facility of historic meaning to the world. From what is learned here, as this partnership between the Atomic Energy Commission and the Duquesne Light Company progresses, will represent a priceless contribution to the harnessing of nuclear power for the benefit of humanity. For Westinghouse, participation in designing and developing the nuclear reactor has been a most challenging and inspiring experience.

The day that the energy was synchronised with the Duquesne Light Company system may have been momentous for the industry, but it passed pretty unremarked by the general public around the Pittsburgh area who were making their toast and brewing their coffee through electric energy generated by the atom.[55] They found out a few days later, with newspaper headlines proclaiming: 'We Enter the Age of the Peaceful Atom.'[56]

Over the next few years other commercial energy stations opened, with the companies favouring different reactor technologies. General Electric, Westinghouse's main competitor, opened the Dresden Nuclear Power Plant. Unlike Shippingport, this was completely privately owned, the first in the country to go on line.[57] And it was also a boiling water reactor

(BWR) – this, like Shippingport, used water as a coolant and moderator, but unlike in a PWR the water is allowed to boil directly in the reactor core, which produces steam that is used to drive the turbine and generate electricity. Dresden I, the first of three reactors on site, started producing energy in April 1960.

Westinghouse and General Electric were the two leaders in the initial nuclear industry, but there were other rivals, like Babcock and Wilcox, and Combustion Engineering. Both Westinghouse and General Electric had gained nuclear experience through government contracts during the Second World War, but General Electric edged ahead of the competition through strength of numbers and money, assigning 14,000 of their work force to nuclear projects and committing $20 million (more than any other company) to research and development.[58]

While the PDRP had encouraged private investment in the emerging technology, it wasn't the only measure that was needed. A crucial component was the passing of the 1957 Price–Anderson Nuclear Industries Indemnity Act, a key piece of legislation enacted by the United States Congress to address the liability and insurance challenges associated with the emerging technology. The Act established a comprehensive system to ensure that adequate compensation would be available in the event of a nuclear accident. It provided a framework for limiting the liability of nuclear operators while guaranteeing those affected would receive fair compensation. The Act also required plant operators to contribute to a common insurance fund, creating a collective pool of resources to cover any potential damages.

A regulatory framework was also established by the AEC that attempted to strike the difficult balance between ensuring that the technology was safe but also not so strict that it prevented growth.[59] Willard Libby, one of the AEC Commissioners, outlined the convictions of the time:

> Regardless of location and isolation, there is no such thing as an absolutely safe reactor – just as there is no such thing as an absolutely safe chemical plant or oil refinery ... Our safety philosophy assumes that the potential danger from an operating atomic reactor is very great and that the ultimate safety of the public is dependent upon three factors: (1) recognizing all possible accidents which would release unsafe amounts of radioactive materials; (2) designing and operating the reactor in such a way that the probability of such accidents is reduced to an acceptable minimum; (3) by appropriate combination of containment and isolation, protecting the public from the consequences of such an accident, should it occur.[60]

Shippingport and the other demonstration reactors were important because they showed the feasibility and economics of nuclear energy, as well as revealing areas that needed improvement and innovation.[61] And while Shippingport played a vital role in building confidence, demonstrations, exhibitions, fairs and expositions were also important platforms for showcasing this new technology.

This method of presenting technological innovations has a rich history. For example, during the late-nineteenth and early-twentieth centuries, the public had mixed reactions to electricity as a new energy medium. While some saw its potential for transforming their lives, others were concerned about its dangers. People feared the power and unpredictability of electricity, worrying that it could harm their health and safety. Some even feared that electric lights could cause blindness or insanity, while others worried that electric currents could disrupt the natural balance of the body and lead to illness or death. High-profile accidents involving electric shock, fire and electrocution, as well as the use of electric chairs for capital punishment, added to these anxieties.

Despite these fears, electricity gradually gained acceptance, largely through familiarity. It was prominently displayed at exhibitions, expositions and world fairs, where manufacturers showcased electrical appliances. Popular attractions were electric restaurants, where all the food was cooked using electricity, or 'all electric' model homes.[62] These fully functional homes demonstrated the latest technologies, including incandescent lightbulbs and ingenious devices that enhanced domestic comfort, like knife-cleaning mechanisms, pianos and bed-warming pans. Visitors were particularly drawn to the kitchens, which showcased the latest stoves, ovens and appliances.

With advancements in technology and the establishment of safety standards, the advantages of electricity became evident to the public. People started to acknowledge its capacity to enhance productivity, improve lighting conditions, offer greater convenience in their daily lives and to rationalise their fears. When it came to promoting electricity generated through nuclear energy, power companies adopted a similar approach of demonstration and familiarisation in the hope of building the same acceptance among the public. There were many examples of this in the US, Europe and further afield, such as small exhibitions like the 1947 Atomic Energy Train Exhibition, which was viewed by 146,000 visitors in cities across Britain.[63] And the 1949 British Industries Fair, at the Olympia exhibition centre in London, where the Ministry of Supply had a stand called 'Atomic Energy in Medicine & Industry'.[64]

At the 1955 International Congress on Peaceful Uses of Atomic Energy, a UN-sponsored conference held in Geneva,

Switzerland, one of the main US exhibits was a nuclear reactor. This reactor had been built at Oak Ridge National Laboratory, dismantled and then transported to Switzerland by plane. It was reassembled during the conference and then brought to full power. Housed in its own building on the grounds of the Palais des Nations, not only was it the first time that people would have seen a working nuclear reactor, it was also possible to view an effect called Cherenkov radiation. Newspapers and magazines around the world reported on this magical blue light that could be seen down the bottom of the reactor's deep cylinder of water-shielded uranium fuel elements. The effect, which had been first observed in 1951 at the Low Intensity Test Reactor at ORNL, is found when electrically charged particles, such as protons or electrons, are moving at speeds faster than that of light in a clear medium, like water or air. When the molecules of the, in this case, water and the charged particles interact, it gives off a beautiful blue glow.

But the demonstration of the peaceful uses of the atom that probably had the longest-lasting reach, in that it still exists today, was to be found at the Brussels World's Fair, an event that was held from 17 April to 19 October 1958. It marked the first major international exhibition since the beginning of the Second World War. Spanning across a vast area just under 500 acres on the Heysel Plateau, the fair focused primarily on the peaceful applications of nuclear energy. Its slogan, 'A world for a better life for mankind', emphasised the belief in technological and scientific progress.[65] The fair garnered immense public interest, with a recorded total of over 41 million admissions, and 48 nations taking part, their exhibitions housed in various national and commercial pavilions.

The most stunning part of Expo 58 was an impressive architectural structure standing at a height of 102 metres and weighing 2,400 tons – all of which was supported by three massive bipods. Known as the Atomium, the structure featured nine interconnected large spheres which symbolised an iron crystal 'magnified 165 billion times'.

Inside these orbs there were displays on peaceful uses of nuclear energy from a number of groups, including the European Atomic Energy Community (Euratom) and Union Minière du Haut Katanga, which showcased a piece of uranium ore covered by a transparent sphere.[66]

As part of the drive to normalise nuclear technology an 'electric house' was also constructed on site, showcasing what could be achieved with 'electric power, drawn from the nearest atomic power station', including heated floors providing 'whatever warmth is required in any area of the house. Air conditioning sees to the mechanical extraction of all smells and ensures that dust infiltration is practically non existent'.[67]

In an effort to familiarise the public with the technology, many nuclear energy plants also offered memorabilia – either souvenirs or promotional. We can see this today through badges for sale, postcards, stamps, ash trays and all sorts of other ephemera. So, the Big Rock Nuclear Power Plant near Charlevoix, Michigan, distributed a postcard proclaiming that, as the largest power plant in Northern Michigan, it will 'provide 50,000 kilowatts on completion'. The Pennsylvania Power & Light (PPL) Susquehanna Nuclear Power Plant also did a solid job of communicating the benefits of nuclear power. Their promotional material, which was distributed by the joint owners PPL and Allegheny Electric Cooperative, was a card with an affixed plastic pouch containing a simulated half-inch nuclear fuel pellet. The text read: 'One nuclear fuel pellet – the size of this simulated pellet contains energy equivalent to about 1,700 pounds of coal or 150 gallons of oil or 160 gallons of regular gasoline. The two reactors will contain 34 million pellets.'[68]

You could also bring a reactor site into your home in the form of the Revell Westinghouse Atomic Power Plant diorama. This intricately detailed educational kit showcased a standard Westinghouse nuclear plant which 'Changes Nuclear Energy into Electric Power'. The kit contained a nuclear-reactor vessel with a removable dome, movable gantry with operational hoist, circulating pumps, steam generator, turbine, electrical substation, high tension tower and separate power lines, service truck, crew, sign, border fence and much more. The parts were cleverly moulded in different colours to minimise the need for painting. It also came with a booklet by a Dr William E. Shoupp – who had worked on *Nautilus* and Shippingport before becoming technical director of Westinghouse's Atomic Power Division – called *A New World of Atomic Power*, which included full instructions. Originally introduced in 1959, the diorama was reissued in 1960 due to its popularity, and is now highly sought after and commands high prices in the collectors' market.

By the early 1960s there were three reactors operating in the US – one government owned and two privately owned – with twelve other units in various stages of planning or construction.[69] The AEC was also aiding the development of no fewer than eleven different types of reactors, with pressurised water and boiling water designs the furthest advanced.[70]

Each nuclear plant typically housed more than one reactor, but even within a single plant, they varied in their specifications, and with no established norms or proven designs. Indeed there was no standardised nuclear reactor readily available.[71] Crucially, as unstandardised and experimental technology, they were not cost competitive with older technologies, such as coal-powered plants, and it remained difficult to convince utilities that they could ever be so.[72]

Science magazine in a 1962 article, 'Atomic Power: Cinderella is Slipping Back into the Kitchen', expressed much of the frustration felt by government and the AEC specifically:

What has happened, most briefly, is that the glamour has gone out of atomic power. Space has taken over most of the position that atomic energy so recently held as a field to be pursued, quite aside from its intrinsic value, as a symbol of national prestige and technological supremacy. Accordingly, the goal of economically competitive nuclear power, once talked about almost in the way the race to the moon is discussed now, has lost much of its sense of urgency.[73]

Glenn Seaborg, who became chair of the AEC in 1961, was committed to accelerating the nuclear energy programme and, in an attempt to mobilise support, convinced President Kennedy to request an investigation that would take 'a new and hard look at the role of nuclear power'.[74] The eventual findings, 'Civilian Nuclear Power ... A Report to the President', was submitted in 1962. The AEC's assessment highlighted the perceived importance of nuclear in meeting the long-term energy needs of the United States, indicating that it was 'clearly in the short- and long-term national interest and should be vigorously pursued'.[75] The report even predicted that nuclear power could potentially contribute up to 50 per cent of the country's electrical generating capacity by the year 2000.[76]

Of significant note was their conclusion that nuclear energy was on the verge of achieving economic viability and could soon 'be made competitive in areas consuming a significant fraction of the nation's electrical energy'.[77] While the feasibility of these analyses and projections were arguably disputable, they played a crucial role in instilling confidence among the public and government officials about the economic potential for civilian nuclear energy,[78] and they led to the construction of more power plants.

On 26 September 1963 there was a ground-breaking ceremony for the Hanford facility's ninth and final reactor, and a
speech by President John F. Kennedy to mark the occasion. This
event took place just over a month after the signing of the Partial
Test Ban Treaty, a tentative halt to the arms race, which pledged
'to end nuclear explosions in the atmosphere, outer space and
underwater'.[79] This agreement not only promised to help contain atomic fallout, but was also expected to redirect the focus
of the AEC towards peaceful applications of nuclear power.[80]

President Kennedy's visit marked the commencement of plutonium production at the N-Reactor and the initiation of the
construction of its power-generating component.

During the scorching heat of the day, approximately 1,500
dignitaries and over 30,000 individuals, including school children from various areas, gathered to witness the ceremony, along
with a number of high-school bands providing entertainment for
the enthusiastic crowds.[81] The event held particular significance,
as it marked a rare departure from the very stringent security
measures that were typically enforced at Hanford.

In his twelve-minute speech Kennedy referred to the atomic
age as a 'dreadful age' with no certainty that humanity could
control the weapon that had been unleashed. More positively
he also reflected that they had the opportunity to 'put science
to work, science to work in improving our environment and
making this country a better place to live'.[82]

After the President's speech, Gerald Tape, the AEC
Commissioner, handed Kennedy an atomic wand measuring eighteen inches in length. With a lovely attention to detail, attached
to the tip, enclosed in a transparent case, was a piece of uranium
which was said to have been taken from the inaugural reactor at
Hanford. As he handed the device over, Tape said: 'Mr President, I
think it is indeed fitting that the breaking of ground for this particular power facility should be initiated through the use of the atom.'[83]

Kennedy waved the wand over a Geiger counter, which started a crane scoop that dumped a load of dirt on the ground. Kennedy seemed interested by the whole spectacle but expressed a slight doubt in the remote-control technology: 'It's a great pleasure to do this ... I assume this is wholly on the level and there is no one over there working it.'[84] Just two months after breaking the ground at Hanford, President Kennedy would be assassinated on 22 November.

While the hope that nuclear energy would provide half of the nation's electricity by the turn of the century turned out to be wildly optimistic, the speech did reflect a growing importance placed on the need for increasing electric energy.

General Electric and Westinghouse, along with some other utility companies, had concluded that a second generation of nuclear plants that applied economies of scale could be competitive with fossil fuels, especially in places where fuel costs were high.[85] By 1963 they launched a series of 'turnkey' projects, in which they assumed all risks in building and offered them to utilities to simply take over. It was hoped that declining costs would help make nuclear energy profitable, even if it meant losing money in the short term.[86]

One of the largest of these projects was the energy plant at Oyster Creek, New Jersey. After General Electric won a bidding war, they were contracted to build a large unit for Jersey Central Power and Light Company.[87] While they expected to lose money on the project, they also wanted to stimulate the market and produce electricity cheaply to make it cost effective. The contract was announced on 12 December 1963 and inaugurated a period that became known as the 'great bandwagon market' as companies jumped on board.[88]

A further boon was created in August of 1964 when the Private Ownership of Special Nuclear Materials Act was enacted by Congress. The Act aimed to facilitate the building of nuclear plants and establish agreements between the AEC and private energy companies. It granted the AEC the authority to offer uranium-enrichment services, known as 'toll enrichment', to domestic and foreign customers through long-term contracts. This allowed them to provide enriched uranium to purchasers, ensuring reliable access to diffusion plants for American nuclear vendors selling reactors domestically and internationally. The Act repealed the mandatory government ownership and monopoly over uranium established by the AEA of 1946, providing the nuclear industry with more flexibility. A multinational uranium market was created with centralised trading agencies, brokerages, lobbying associations and cartels.[89] While there was a transition period, by 20 June 1973, private ownership of reactor fuels became mandatory, ending the need to lease from the government.[90]

While this was great news for the nuclear energy industry, it did have wider implications and highlighted the need for extra safeguards against the diversion of special nuclear materials and the possibility of more nations obtaining atomic weapons. The AEC extended their controls to privately owned materials, introduced improved bookkeeping and inventory procedures, and initiated further study on the necessity of implementing physical security measures.[91]

Power companies turned to nuclear energy for electricity generation, leading to the operation, construction and planning of numerous new energy plants.[92] By August 1966 there were fifteen nuclear energy plants in operation, eight under construction and twenty on the drawing board.[93]

With a growing number of nuclear power plants there was a corresponding need for more uranium. But the problem was that again, there was a shortage. What later became known as the first uranium boom had ended in October 1957 when the AEC announced: 'We have arrived at the point where it is no longer in the interest of the government to expand the production of uranium concentrate.'[94]

In the late 1950s the demand for uranium from the burgeoning nuclear industry was simply not enough to maintain the industry at previous levels. And with more than enough uranium for their needs and a stockpile of over 30,000 strategic nuclear warheads, the AEC had phased out their previous policies that encouraged new discoveries and limited purchases to already established mines.[95] This had a devastating effect on prospective uraniumaires, some of whom were in the process of registering their mines as the announcement was made.

Without the same level of demand, many mines were simply abandoned, with the inevitable loss of livelihood for workers. Towns, which had seen their populations grow, also found themselves in decline or were abandoned altogether. And this was a significant problem for many places – there had been over 7,500 reports of uranium discoveries in the US, with 850 underground and 200 open pit mines producing uranium.[96] Additionally, there were 27 uranium processing mills in operation by 1960, spanning from Washington to Texas.[97]

Now as a new, albeit much smaller than previously, mining boom took hold in the mid-1960s, the AEC announced substantial investments in drilling by 49 companies during a three-year period, amounting to $77 million.[98] However, the second uranium boom had a very different feel from the first. It was no longer possible for individual prospectors to reap the benefits and become wealthy. Those smaller companies were effectively pushed out and replaced by large corporations like Exxon, Union Carbide, Kerr-McGee and Getty Oil, who increased their already established presence in the area. Time had moved on from the 'scores of speculative little companies financed by penny stock issues and hundreds of individual prospectors' as surface uranium deposits became scarcer, requiring deep drilling equipment and therefore greater investment.[99]

In practice this decimated the dreams of countless independent prospectors, many of whom still harboured the uranium bug. Old hand Charlie Steen expressed his concerns: 'Under present conditions, I would hate to be the prospector who finds another Mi Vida … Anybody who goes out and prospects for uranium now is a damned fool.'[100]

The second uranium boom posed even greater challenges for mine workers compared to the first. One significant distinction was the shift from working in shallow pits to deep pits, which introduced new risks. Moreover, safety standards in the 1960s differed significantly from those in the previous decade. Lessons learned triggered increased scrutiny, prompting more people to examine the realities and health risks of working in uranium mines.[101] It was the first-hand testimonies of individuals engaged in this work that truly shocked many, particularly the lack of adequate protection against exposure to radioactivity.

Mining of any kind is always dangerous and is characterised by gruelling labour, inherent hazards and detrimental health effects. But uranium mining had its own set of hazards. As far back as the sixteenth century, the fate of the miners of Erzgebirge in Europe, with their unusually high rates of respiratory illnesses and fatal lung diseases, was widely known, with Joachimsthal's town physician Georg Bauer reporting: 'It eats away the lungs, and implants consumption in the body, hence in the mines of the Carpathian Mountains, women are found who have married seven husbands, all of whom this terrible consumption has carried off to a premature death.'[102]

That what he was describing was lung cancer (and nothing to do with mountain gnomes seeking revenge) had been first officially acknowledged in the late 1870s. That it was actually caused by the products of uranium decay chain was first proposed in 1913. But while the Czechoslovakian government had taken steps during the 1930s to reduce the danger by installing mine ventilation, in the US there was significant opposition to the argument that radon in the quantities that were being measured in the mines was a health risk.[103] Though the dangers of radiation were known, it was argued that the concentrations were so small that their exposure would also be equally low.[104]

It took a decade of warnings and surveys into the conditions faced by the uranium miners before the US Public Health Service (USPHS) released their findings and attention came to be focused on the issue. The USPHS reports spurred some mine owners to improve their ventilation and to establish minimum standards for radon concentrations, but this was largely inadequate. There was a resistance to doing any more and a lack of clear protocol over which authority had responsibility for mine safety in regards to radiation.

This reluctance was compound by the fact that although the USPHS had the authority to make recommendations, they did not have the power to shut down mines for non-compliance. Furthermore, they had agreed not to warn the miners directly about radiation hazards or let them know about the findings of their health studies.[105] Their conclusion was chilling, noting that while they hadn't found radon-related cancers in the miners, it was 'not entirely surprising', as the latency period for the disease had not yet passed.[106]

By the late 1950s the USPHS studies had been expanded to include even more miners and had gained interest from the military, who had wanted to conduct studies into the effects of moderate doses of radiation but had found the ethics of this type of study difficult to side-step. The opportunity was noted:

'Advantage should be taken of any opportunities for the study of the biological effects of radiation, particularly in man.'[107]

The *Washington Post* played a vital role in exposing the ineffectiveness of federal agencies in regulating the mines by publishing a series of impactful articles and editorials with hard-hitting headlines like 'AEC Death Mines', which brought the issue to public attention.[108]

While the AEC disputed that characterisation of their position, arguing that during the first uranium boom they had made many advances in mine safety and maintaining the health of the workers, they also approached the other agencies involved in the industry to try to come to a joint standards programme. They failed to reach a consensus, leading to a deadlock.[109]

Eventually, under the Department of Labor's authority, a radon standard, expressed as 3.6 'working level months' (WLM) per year was suggested, triggering debates and criticism from various stakeholders questioning the scientific data behind these calculations and pointing to the detrimental impact that it would have on the industry, in effect forcing many more mines to close. President Johnson approved the eventual recommendation, of a twelve WLM limit, in 1967. While it was largely viewed as a step in the right direction, there were still some fierce critics, with the President of the United Mineworkers of America commenting that the deaths of uranium miners from lung cancer 'underscores once again the brutal cost that this nation has paid to develop the atomic industry'.[110]

The nuclear energy turnkey programme ended in 1965, and while it had cost both General Electric and Westinghouse many hundreds of millions of dollars, as expected, it had convinced several utilities in high fuel-cost areas that nuclear energy was a sound investment.[111]

And it was an industry eager to produce electricity for a 'High Energy Society', characterised by a power-hungry public with an abundance of modern appliances that thought of itself as 'a truly modern technological culture whose measure of advancement can almost be equated to its consumption of energy – and particularly energy in its most usable form: electricity.'

This was also a period when earlier concerns about fallout had largely disappeared, opponents of nuclear energy were few, capital costs were low and press reports were largely positive.[112] Public sentiment towards the technology was also predominantly positive. Opinion polls conducted at the time revealed that 69 per cent of the US population expressed no fear if a nuclear power plant were to be built in their area, while only 20 per cent reported feeling any apprehension to the thought. Fossil fuels were more likely to be thought of as the danger, with their devastating effect on air quality, and in contrast nuclear energy was seen as 'safe, clean, quiet and odorless'.[113]

It was argued by many – including environmental groups – that nuclear energy, especially with its small construction footprint, could play a major role in improving standards of living throughout the world.[114] As William Siri, President of the conservation organisation the Sierra Club, which once had the slogan 'Atoms not Dams' predicted:

[Nuclear power] is one of the chief long term hopes for conservation, perhaps, next to population control in importance ... Cheap energy in unlimited quantities is one of the chief factors in allowing a large rapidly growing population to preserve wildlands, open space and lands of high scenic value ... [The discussion on nuclear power] will not end until some future date when our children look back from the clean comfortable

world driven by nuclear energy and wonder what all the fuss
was about. The rest of the universe runs on nuclear energy,
why not us.[115]

However, as the 1960s progressed, anti-nuclear energy movements gained momentum, which had their origins in protests
against nuclear armaments and testing. Initially, opposition
was not against the ideology of nuclear energy, but, instead,
objections focused on specific proposed sites and therefore was
largely regional in nature.[116]

One notable case was the planned site at Bodega Bay,
California, for Pacific Gas & Electric (PG&E). Local residents
expressed concerns about the visual impact of the plant, its
effect on property prices and the potential harm it could cause
to the local fishing industry. When these concerns failed to
sway the AEC away from licensing the site, the opposition
shifted its focus to the potential seismic hazards of the location. Although the planned plant would be situated near the
San Andreas Fault, PG&E conducted research and concluded
that there were no major faultlines on the actual site, while also
asserting that the design could withstand any earthquake up
to 8.2 on the Richter scale. The concerned residents, who had
organised into the Northern California Association to Preserve
Bodega Head, commissioned research of their own which came
to an opposing conclusion. The arguments culminated in a large
public meeting in June 1963 in protest of the power plant. Here,
1,500 helium balloons were released into the sky carrying the
rather over-egged message: 'This balloon could represent a radioactive molecule of strontium 90 or iodine 131. PG&E hopes
to build a nuclear plant at this spot, close to the world's biggest
active earthquake fault. Tell your local newspaper where you
found this balloon.'[117] PG&E eventually withdrew their application and the plant was never built.

As the nuclear energy industry continued to grow, so did the opposition against it. And over time the localised campaigns drew attention to broader challenges.[118] The turning point was concerns raised around a debate over thermal pollution in the late 1960s. The issue was the water used for cooling the steam that powered the electricity-generating turbines. This water was drawn from nearby sources and returned to the same location, albeit at elevated temperatures.[119] The issue became a flash point, with concerns around the ecological impact of these warmer, oxygen-depleted waters increasing.[120]

The AEC initially refused to address the issue, asserting that, legally, it fell outside of their jurisdiction. The controversy spilled into the public arena and was seized upon as a key campaign area by environmental groups despite the power companies trying to positively rebrand it as 'thermal enrichment'.[121]

The issue of the warmer water was easily solved – although with additional costs – by the construction of cooling towers, used for water evaporation, to remove process heat. The unclear jurisdictional authority was eventually addressed through federal legislation. However, the resolution of this crisis came at another cost to the nuclear energy industry, as it experienced a significant loss of authority in the process. The lack of clarity and the controversy surrounding the AEC's handling of the issue further eroded public trust and confidence in the technology.

The end of the decade saw a clash on several fronts, including a heated corporate battle, with the coal industry fiercely competing against the nuclear industry. While the two types of energy producers had largely adhered to an earlier non-official agreement not to compete head to head, an 'Atomic Bomb in the Land of Coal' was set off when the Tennessee Valley Authority, a coal-mining region, chose to build two General Electric reactors at Browns Ferry, Alabama, instead of a coal-fired plant.[122]

In retaliation the coal industry began to use their significant resources to actively oppose any new nuclear plants.[123] When utilities contemplated constructing nuclear energy units, coal companies responded by significantly lowering their fuel prices or offering price concessions. This strategy aimed to reduce the overall cost of power production and maintain the competitiveness of coal as an energy source, which itself had benefited from decades of subsidies and investments.[124]

The larger nuclear industry, and specifically the AEC, also failed to quickly adjust to the changing priorities of the 1960s, specifically in terms of protecting the environment. When the Baltimore Gas & Electric Company proposed building the Calvert Cliffs Nuclear Power Plant, Maryland, nearby residents asked that the thermal pollution should be taken into account and eventually took the AEC to court. The court's decision, which ruled that the AEC was not compliant with the recently enacted National Environmental Policy Act, had far-reaching implications for future licensing procedures. As a result, the court mandated that the AEC undertake an independent review and evaluation of all environmental effects at every stage of the nuclear energy plant licensing process.

This decision triggered a significant restructuring of the AEC's licensing procedures.[125] Both the AEC and the licence applicants were now required to consider the overall impact of the proposed plant on the environment, including factors such as water quality. Additionally, a cost-benefit analysis was to be conducted, weighing the advantages of constructing the facility against various alternative options.[126] These procedural changes affected virtually all nuclear energy plants, whether they were already licensed for operation or still under review. While important, these changes also ultimately increased both the time and costs required to bring these plants into operation.[127]

To compound the fall from grace by the end of the decade, there was also a noticeable change in the press's stance on nuclear energy, which had previously been largely supportive.[128] Publications such as Sheldon Novick's *The Careless Atom* played a role in shaping this change. The book not only highlighted concerns about the technology, but also drew national attention to the controversial issues surrounding the proposed Bodega Bay nuclear energy plant. Worst still for the industry were the marketing decisions made by the book's publishers, Houghton Mifflin, who used attention-grabbing and ominous taglines, such as 'The Hiroshima Bomb is alive – and ticking – in Indian Point', referring to a nuclear energy plant located in New York. This wasn't a one-off, either, with the publishing house Doubleday marketing another anti-nuclear book, in July 1969. *Perils of the Peaceful Atom*, was released with an advert stating: 'The "peaceful uses" of atomic energy can kill you just as dead.'[129]

While supporters of nuclear complained that these books contained many glaring errors, exaggerated depictions of risk and a lack of understanding of the technology – especially the popular misconception that a reactor could blow up like a bomb – the marketing tactics of these publishers only further contributed to the growing scepticism and apprehension surrounding nuclear energy.[130]

There was, however, a potential lifeline for the industry. As the 1970s started there was also a growing apprehension about the availability of electricity in general, with warnings of an impending crisis looming on the horizon. It was projected that a shortage of electricity would likely occur in the 1980s due to increased usage levels. However, the crisis manifested earlier than anticipated, as demand surged at a faster rate than expected.

There was also the 1973 oil crisis, when members of the Organisation of the Petroleum Exporting Countries (OPEC) organised an embargo against nations that had supported Israel in its recent conflicts, including the US and the UK. This caused supplies to dwindle, the cost of oil to rise by around 300 per cent and led to an internal crisis as industries had to shut down because of lack of fuel while the public literally fought over the decreasing amounts available.[131] Rising oil prices led to price increases in all commodities, as well as the cost of energy from all sources.

The urgency to address the growing energy demand and pursue alternative power sources led to a renewed focus on nuclear, and it became a central component of President Nixon's 'Project Independence'. This initiative aimed to achieve self-sufficiency for the United States by 1980 and included emergency powers, relaxed environmental standards and increased authority for the AEC to issue temporary operating licences without public hearings.

However, the conflict between energy companies and environmentalists hindered the consensus on the President's proposals. Groups like Friends of the Earth, who spearheaded the opposition against nuclear energy in the early 1970s, viewed the technology as a 'threat to the safety of life on Earth' and called for a moratorium on the construction and operation of all nuclear plants. They accused President Nixon of capitalising on the energy crisis to bypass environmental safeguards and accelerate the development of nuclear energy.[132] Two years later, when Nixon was compelled to resign from office after revelations about his abuse of power and corruption, the project faced even more criticism and opposition from various interest groups.

Despite all of this, opinion polls consistently indicated that the public still held favourable views towards nuclear electricity, with substantial margins. In a 1969 survey, 50 per cent of respondents expressed support for nuclear energy in their local area, while 27 per cent opposed it.[133] And the utility companies still demonstrated their confidence through orders for new builds. By the end of 1974, there were 233 nuclear generating units in the United States, either in operation, under construction or on order.[134] Worldwide, there were nineteen countries that had added nuclear energy to their mix, with a total of 157 plants in operation and more in the works.[135]

On 11 October 1974, the new president of the United States, Gerald Ford, signed the Energy Reorganization Act, marking a significant change in the energy landscape. This act formally abolished the AEC and created two new agencies: the Nuclear Regulatory Commission (NRC) and the Energy Research and Development Administration (ERDA), taking over the regulatory and licensing functions, and the research and development responsibilities respectively.[136] Three years later, the ERDA was dismantled, and its duties transferred to the newly formed Department of Energy. The NRC was periodically restructured but ultimately suffered due to its dual role of promoting and regulating the industry, which often brought it into conflict with itself.

The late 1970s also marked a significant turning point for the nuclear energy industry, as new plant orders diminished, largely due to an overestimation of demand and steadily increasing capital costs, and local opposition to construction projects grew, largely driven by the environmental movement. By the time Jimmy Carter assumed the presidency in 1977, new orders for nuclear generating equipment in the US had virtually ceased, and even ongoing construction projects encountered difficulties.[137]

The stagnation in nuclear energy development had a profound impact on the energy landscape. Prior to the slowdown, nuclear was responsible for supplying around 20 per cent of the country's electrical power. However, with the halt in new construction and the absence of further expansion, growth came to a standstill. The industry entered a period of dormancy, and its contribution to the nation's power supply remained at the existing level. As ever, other forms of power generation were poised to take over, with some sites, which had only been at the planning stage, being converted to coal.[138]

At the same time, environmental concerns experienced a significant surge, particularly in highly industrialised countries. This led to the emergence of organisations that swiftly turned their focus towards opposing nuclear of any kind, citing environmental protection issues, used-fuel disposal and concerns around radiation.

In 1971, the Committee for Nuclear Responsibility became the first national anti-nuclear group, demanding a halt to the construction of all new nuclear plants.[139] The Sierra Club, previously a supporter of nuclear energy on environmental grounds, shifted its stance and began actively opposing it, prioritising other energy sources, including coal, instead. Their approach involved legal action, and by the early 1970s, they were involved in 55 lawsuits across the country.[140] While the early years of nuclear power plant construction had been largely without citizen opposition, by 1970, 73 per cent of all licensing applications were challenged by various organisations.[141]

While Friends of the Earth and the Sierra Club called for a moratorium on nuclear power development, other groups focused on protesting individual projects. The movement was characterised by a variety of tactics, including lobbying and petitions to direct actions and civil disobedience.[142]

Countries prioritising nuclear development were met with fierce resistance. Starting in the early 1970s, France built 56 reactors in just fifteen years as part of the Messmer Plan. Introduced by Prime Minister Pierre Messmer, the aim was to ensure that 75 per cent of the country's electricity was generated by nuclear reactors. This major push was characterised by continuous domestic protests and incidents of low-level sabotage from the opposition movement against nuclear energy, which included Les Verts, the French political party focused on environmental issues, and Sortir du nucléaire, a national network advocating for the phase-out of nuclear power, represented a significant number of individuals and organisations opposed to nuclear energy.[143] In July 1977, a large demonstration took place with the aim of halting the construction of the Superphénix, a nuclear plant at Creys-Malville near the France–Switzerland border. Approximately 60,000 people participated in the protest, which unfortunately resulted in serious injuries and one fatality.[144]

At Wyhl in West Germany, a 1975 protest saw the occupation of a proposed reactor site followed by a demonstration consisting of more than 20,000 people. The site was never built.

Taking inspiration from the Wyhl protests, the Clamshell Alliance was formed in New Hampshire in 1976 to oppose the construction of the Seabrook Station Nuclear Power Plant. They organised two small occupations, followed by a large-scale occupation involving 24,000 people in the spring of 1977, during which 1,400 were arrested.[145]

That this was now a worldwide movement was highlighted by a badge with a 'smiling sun' design created by Danish activist Anne Lund in the mid-1970s. This internationally recognised trademark features a red sun at the centre with the words 'Nuclear Power? No Thanks' written around the edges. The symbol has been produced in 45 different languages and has served as a powerful visual statement against the use of nuclear energy worldwide.[146]

It was clear that the popularity of nuclear energy was on the wane and, as the 1970s drew to a close, one major event dealt a crucial blow to the atomic dream and would effectively end the first hopeful history of uranium.

And on 28 March 1979, at the Three Mile Island (TMI) Nuclear Generating Station on the banks of the Susquehanna River near Harrisburg, Pennsylvania, a combination of equipment malfunctions, design-related issues and operator errors led to the reactor core of Unit 2 partially melting down, which resulted in the release of a small amount of radioactive material and a large amount of international condemnation.[147]

NUCLEAR NIGHTMARES AND NUCLEAR DREAMS

8

It's interesting that an incident with no immediate injuries, deaths or direct health effects and no significant environmental damage can essentially ruin an industry.

But that's what happened.

I made the decision to end this book in 1979 not because it was the end of the story – a heavily subsidised business doesn't die as easily as that – but because it was the point at which the narrative of uranium became damaged seemingly beyond the point of repair.[1]

The accident was undoubtedly a significant event that caused widespread concern and panic at the time and understandable post-traumatic stress disorder afterwards. Although we now know that, thanks to the containment structure doing its job, there were no long-term consequences, the lack of available information during the unfolding situation contributed to the heightened levels of anxiety.[2]

The response from both the government and the nuclear industry was not well coordinated, creating a climate of uncertainty and leading to further confusion and fear. Nearly 150,000 people, dubbed by the press as 'nuclear refugees' or 'nuclear evacuees', fled their homes voluntarily before returning ten days later.[3] The mental stress experienced by residents was substantial, as letters received by Harold Denton, appointed to act as the personal representative of President Carter, reveal. As one person wrote: 'I did not evacuate the area during the recent crisis, but I did help friends and relatives do so. Needless to say, the toll in human suffering was great, mainly in terms of the psychological stress generated by our loss of control over our own lives.'[4]

And an estimated 400 reporters descended upon the scene, seeking to provide first-hand coverage and updates on the unfolding events. The deluge of reporting created a saturation of information, with broadcasts airing extensive segments and some newspapers publishing articles with ominous (and palpably untrue) headlines: 'TMI: Terror Unleashed',[5] 'Nuke Cloud Spreading', 'Race with Nuclear Disaster' and 'Life'll Never Be Same, Nuclear Refugee Says'. Concerns were also raised about the potential of a hydrogen bubble forming in the damaged reactor, which could potentially explode at any time. While this was never really a possibility, it helped to firmly establish the conflation of nuclear power and nuclear war, a propaganda that the fossil-fuel industry and anti-nuclear institutions were keen to emphasise.

In addition, the unfortunate timing of the newly released disaster film *The China Syndrome* intertwined with the unfolding situation, amplifying the public's perception of the risks and potential consequences associated with nuclear energy. Starring Jane Fonda, Jack Lemmon and Michael Douglas, the film depicted a fictional reactor crisis, a cover-up and a group of individuals determined to reveal the truth, and was released just twelve days before the incident.[6] The coincidence was seemingly perfect for media exploitation, with *The New York Times* headlining a piece: 'When Nuclear Crisis Imitates a Film', and *Variety* hailing Three Mile Island a 'powerful trailer' for the film.[7]

A quote from the movie was widely circulated afterwards as if it was a prophecy: 'If the core is exposed, for whatever reason, the fuel heats beyond core heat tolerance ... Nothing can stop it, and it melts right through the bottom of the plant, theoretically to China. The number of people killed would depend on which way the wind is blowing [and] render an area the size of Pennsylvania permanently uninhabitable.' Of course, the idea that nuclear fuel could melt its way down to the core of the earth is impossible, but it does make for great cinematic tension.

The impact of the accident also extended beyond the immediate vicinity and spilled over into a national concern. The extensive press coverage, coupled with the widespread propagation of information, contributed to a sense of unease and anxiety among people across the country. It also, of course, did not escape the attention of popular culture and was referenced and depicted in various forms of media, further embedding it in the public consciousness and contributing to broader cultural discourse surrounding nuclear energy.[8]

Just over a week after the incident, the TV show *Saturday Night Live* aired a sketch called 'The Pepsi Syndrome'. At the fictional Two Mile Island plant a clumsy technician, played by Bill Murray, spills a soft drink on the console triggering 'Pepsi Syndrome', which causes a nuclear reactor meltdown, or 'surprise', as they term it. Over the following few days, as the plant's PR team is shown winging into action, they are visited by President Jimmy Carter, portrayed by Dan Aykroyd, who enters the reactor core to assess the situation, gets irradiated and becomes the 'Amazing Colossal President'. It is funny, completely inaccurate and very disingenuous.

While the initial response was mixed, President Carter quickly appointed a commission led by John Kemeny, tasked with investigating the accident. Over the course of six months, they conducted extensive research, which involved taking more than 150 depositions and interviewing numerous individuals under oath.[9]

Based on their findings, the Kemeny Commission made many recommendations for improving safety in the nuclear industry. They suggested restructuring the NRC as well as recommending better equipment in reactor control rooms, investment in training operators and improvements in emergency preparedness at reactor sites.[10]

Consolidating the Kemeny Report, the NRC developed a TMI Action Plan. Covering 347 detailed actions over four areas including operational safety, design and emergency planning, it was structured to serve as a framework for both the NRC and the nuclear industry. The goal was to prevent similar incidents in the future and ensure the adequate protection of public health and safety in the operation of nuclear power plants.[11]

While the regulators and industry were focusing on dealing with the aftermath and addressing safety concerns, the licensing process was put on hold and no licences for new reactors were issued for a year. Many projects that were in the planning or construction phase were suspended or cancelled.[12]

At the same time TMI became emblematic of failure, of the inevitability, as opponents claimed, of nuclear catastrophe. As such it entered the cultural consciousness in a number of ways, including a range of irreverent products. T-shirts with slogans like 'I survived 3 Mile Island, I think' and 'Radiation Doesn't Scare Me, Three Mile Island 1979' next to an image of conjoined twin babies were produced. A sealed can was sold by Brenster Enterprises of Etters, Pennsylvania, which featured a prominent label with an illustration of the power plant and the details of its contents – 'Canned Radiation from TMI,' i.e. it was full of air. The label also humorously suggested six potential uses for the product:

1. Remove label and tell your enemy it's laughing gas.
2. Energy-free night light (illuminates in darkness).
3. Mix with cold cream for that radiant beauty.
4. Instant male sterilization (sniff twice daily).
5. Use as a room air freshener.
6. Toothpaste recipe: mix 3 to 1 ratio with baking soda, for everglowing smile.[13]

Despite the fact that the malfunctioning reactor was actually housed in a containment building nearby, it was the plant's four cooling towers that became the focus of not only jokesters, but also as a way of commemorating and amplifying the incident. A range of very 1970s brown chunky pottery mugs and lamps were created in their shape. There was also a lamp with a shade with a colour print of the towers and on the base an inscription that reads: 'NUCLEAR ACCIDENT, MARCH 28, 1979 MIDDLETOWN, PA.'

The towers were also the illustrative choice for cover stories in *Time* and *Newsweek*, forming nightmarish backgrounds to stories with headlines such as 'Nuclear Nightmare'. This was clearly an editorial choice, as publications opted to use zoomed-in shots of the towers, taking the photographs at either dusk or dawn, which served to give them an eerie and menacing quality.[14] The cooling towers are still standing, although there are plans to dismantle them. The reason for their survival was that, despite plenty of opposition, the undamaged reactor unit 1 was restarted in the middle of the 1980s, operated until 2019 and is now in decommissioning, a process where it is safely dismantled.

This period directly after the Three Mile Island incident was marked by an uptick in demonstrations and increasingly larger and more high-profile public protests. And anti-nuclear sentiment extended beyond the United States; there were protests in many other countries, from the Philippines to Germany, where 100,000 protesters descended on the Brokdorf Nuclear Power Plant. In France a foreign terrorist attack targeted the Superphénix plant, launching five exploding warheads with a Soviet rocket launcher, across the Rhône into the side of the still-under-construction containment building. Though two rockets hit and caused damage to the outer concrete shell, they missed the reactor's empty core.[15] Public acceptance of nuclear energy also dropped significantly, with one poll in the US indicating that only 32 per cent of respondents had any confidence in it.[16] In San Francisco and Philadelphia, 'die ins' were held outside nuclear utility offices. In September 1979, hundreds of thousands of people attended a series of 'No Nukes: The Muse (Musicians United for Safe Energy collective) Concerts for a Non-Nuclear Future' at Madison Square Garden in New York City. With performers such as Bruce Springsteen, Graham Nash and Carly Simon, the concerts were also released as a triple live album and a film to document the performances.[17] Also present was Jane Fonda, a particularly passionate and articulate opponent of nuclear energy – her previous record of activism bolstered by her appearance in *The China Syndrome*. Such was her vehemence that pro-nuclear scientist Edward Teller said after his heart attack: 'you might say that I was the only one whose health was affected by the reactor near Harrisburg. No, that would be wrong. It was not the reactor. It was Jane Fonda. Reactors are not dangerous.'[18]

By the early 1980s, radioactivity had become widely associated with toxicity, from the EPA's campaign against radon gas, deemed a multimillion dollar 'radon fright' train by some in the scientific community, to another round of concerns about the radioactivity of Fiesta Tableware, still in use in many homes despite the fact that they had been discontinued due to low sales in 1972. Incidentally, collecting uranium glass and Fiesta ware is perfectly safe but I would avoid eating and drinking from them – it isn't the radioactivity that causes concern but uranium's properties as a heavy metal, which is toxic and can cause damage to your internal organs. There were also readers' letters in newspapers asking for advice on the safe disposal of radioisotope products, including one known as the Staticmaster. These were brushes, available since the 1950s, that were used to remove dust from film, lenses and records. Made by the Nuclear Products Company of California, they contained radioactive polonium-210 and were advertised as: 'A modern tool for neutralizing static electricity and removing static attracted dust and lint. Brush slowly with radioactive element close to work.' The advice given was to 'send the brush back to the company who will store the old brush in a 50-gallon steel can. When the can is full of brushes it will be buried at a low-level radiation dump.'[19]

While there were concerns about radioactive consumer items, it was the question of nuclear waste that became the biggest flashpoint during this period. The term nuclear waste actually encompasses several different categories of material, from Very Low Level Waste (VLLW), which can be disposed of like any other waste, to Intermediate Level Waste (ILW), which needs special handling. It is estimated that about 97 per cent of waste produced in commercial reactors is either low or intermediate waste. High Level Waste (HLW), a category that includes used fuel rods, remains highly radioactive for extended periods and requires special disposal to ensure safety.

On the big screen, any type of radioactive waste became 'toxic waste' – visually coded as a green gooey substance usually found in rusting barrels. And whereas the majority of irradiated cinematic monsters and superheroes were given their powers through exposure to the effects of atomic testing, by the 1980s the focus is on waste. There are a plethora of films that fit this profile, but to choose just a few of my favourites …

In the comedy film *Modern Problems* (1981), Chevy Chase portrays the character Max Fiedler. One night, while driving behind a nuclear-waste truck, he accidentally gets covered by leaking green goo, which grants Max telekinesis powers. With his newfound abilities, Max aims to win back his love interest and seek revenge on those who wronged him. The film's tagline is explicit: 'Nuclear shower gives him the power.'

The 1984 movie franchise *Toxic Avenger* gives us Melvin Ferd, a bullied janitor who is thrown into a drum of green goo. As a result, he undergoes a transformation, becoming the mutated superhero known as the Toxic Avenger. With a severely injured face and a muscular body, he uses his newfound powers to save the girl and fight against injustice. The franchise incorporates various adaptations, such as comics, musicals and a TV series, as well as further films.

In the same year, *C.H.U.D* (Cannibalistic Humanoid Underground Dwellers or Contamination Hazard Urban Disposal) depicts the Nuclear Regulatory Commission as being responsible for the emergence of monsters lurking beneath the city streets. These creatures are former humans turned into flesh-eating beings due to the radioactive waste illegally transported and concealed by the NRC in abandoned subway tunnels.

While these scenarios are entirely fictional, there was a genuine necessity to address the management of spent nuclear fuel at this time. This need wasn't driven by the fear of barrels leaking in the sewers – even though we are reliably informed that this could lead to the existence of pizza-loving, crime-fighting turtles, which seems rather a good thing. The quantity of spent fuel generated was, and still is, remarkably small. It's estimated that if a person's lifetime of electricity were solely generated from nuclear power, the waste would only fill a soda can, and just a small portion of that would be considered long-lived waste. Coupled with the relatively small quantities produced, the industry is so highly regulated that, unlike other energy producers, they are obliged to account for and responsibly dispose of all waste.

Like thermal pollution in the 1960s, the question of nuclear waste became a key area of campaigning from anti-nuclear groups who identified it as 'the soft underbelly of the industry'.[20] With spent fuel such a hot topic it also became – and indeed still is – fodder for the press and cinema, both of which helped to politicise and magnify the supposed issue.

The question of what to do with used fuel in particular had been under consideration for a long time. A recycling method known as PUREX (Plutonium and Uranium Recovery by Extraction) had been developed in the 1940s. The first commercial reprocessing plant had been built in the 1960s. Countries including France, Russia and Japan had success with reprocessing spent fuel which still contained some uranium-235 and also plutonium that was created during the fission process. This can be recycled as fresh fuel, saving up to 30 per cent of the natural uranium otherwise required. In the UK reprocessing was undertaken at the Windscale site, which was renamed Sellafield in 1981. Here the spent fuel was dissolved in nitric acid and the useful elements, such as uranium and plutonium, separated out from the not so useful ones. From there, highly radioactive wastes can be converted into a glass form via a process known as vitrification. This glass is solid, stable and can be safely stored.

Despite these innovations, in the US technical and political issues had prevented a workable process for the relatively small amounts of high-level waste that were the result of the operation of nuclear energy plants. An operational licence for a reprocessing plant in Barnwell, California, was denied by President Carter, who also, in 1977, halted reprocessing entirely.[21]

Although President Reagan later lifted the ban on commercial reprocessing by 1980, the idea of treating spent fuel in this way was no longer a viable option for many countries as costs skyrocketed.

Without reprocessing, used fuel rods were stored in cooling pools on reactor sites. While this was a safe solution, the problem was that at multiple plants it was only a matter of time before they reached capacity.[22]

There was a proposal for a deep geological facility within the continental United States, which by the end of the 1980s focused on a designated area in the Yucca Mountains, Nevada.

The search for a suitable disposal site for civilian waste would continue for years, involving extensive evaluations, public consultations and legal battles – and indeed it is still a contentious issue, not least because the chosen area is shared by both the Western Shoshone nation and the Goshute tribes. Justifiable concerns about ensuring the safety of long-term storage of used fuel were mixed with worries about nuclear proliferation, with opponents raising scenarios that reprocessing and storage facilities would be at risk of terrorist attack and the material used to make crude nuclear bombs.[23]

While the question of geological storage rumbled on, dry cask storage was identified as the most effective solution to simply keeping the spent fuel rods in cooling ponds for indefinite periods. Once sufficiently cooled, the fuel is transferred into dry casks, typically consisting of a sealed metal cylinder surrounded by a concrete outer shell. These casks not only contain the radiation, but also provide resistance against natural disasters such as earthquakes, floods and extreme temperatures. Their adoption gained momentum, and the first licensed dry storage installation in the United States was established at the Surry Power Station in Virginia in 1986, under the supervision of the regulators. However, even their existence seems to be contentious for many, and despite having a perfect safety record. In 2023, Madison Hilly, founder of the Campaign for a Green

Nuclear Deal, faced criticism for a photoshoot where she pressed her pregnant belly against the side of a dry storage cask at the Idaho National Laboratory. While press coverage was largely even-handed, it was notable the number of memes and comments on social media that fell on those old tropes of genetic deformity and danger. Despite that, Hilly's message largely cut through:

> The trouble with talking about nuclear waste is that most people don't know what it is or what it looks like. Thanks to *The Simpsons*, many people think nuclear waste is a bright green liquid and stored in leaky oil drums ... I find that when people see pictures of nuclear waste and discover that it's actually solid metal, safe enough to hug [when safely stored] and actually quite boring – it alleviates a lot of their anxieties.[24]

While modern commercial spent fuel is perfectly manageable, the legacy of waste from nuclear testing programmes and the Manhattan Project should be a source of deep shame. Of the many atomic sites established during war time with lasting and toxic nuclear heritage is the nuclear weapons production facility known as Rocky Flats, just outside of Denver. Early atomic history was very much characterised by single-mindedness, haste and an almost complete disregard for consequences. Some might say that is the price of war, but the lack of commitment to clearing up after themselves extended into the 1980s, when the nine nuclear reactors on the Hanford site were finally shut down.[25] Today, a group called the Hanford Challenge do amazing work not only to hold those responsible accountable, but to clear the area and promote a sustainable environmental legacy.

Until the signing of the Comprehensive Nuclear-Test-Ban Treaty in 1996, over 2,000 nuclear test explosions were detonated, affecting the environment and the Shoshone, Apache, Pacific Islanders to name a few. That there are different levels of concern depending on which groups are affected is clear. In 1979 there were also two very different nuclear disasters. The first, Three Mile Island, was covered extensively and led to a period of reflection and change. The second was not afforded quite the same level of scandal, although it arguably deserved more.

On 16 July 1979, some 1,100 tons of uranium waste and 94 million gallons of toxic water were unintentionally dumped into the Rio Puerco, New Mexico. The spill was the result of a breach of a holding pond on a uranium mining site managed by the United Nuclear Corporation. The remediation efforts of the UNC were abysmal. They did not seek to communicate the severity or the source of the accident to those who lived nearby and who relied on the river for drinking, irrigation and watering their livestock. The Governor refused to declare it a federal disaster area, despite the requests of the Navajo Nation. And, even after a Congressional hearing into the incident, the company were allowed to reopen the mine and the mill on site, processes that again built up the toxic solution which continued to seep into the river. An inadequate warning sign was put up: 'Contaminated Water Keep Out.'[26]

Against all odds, the nuclear energy sector had actually made a remarkable recovery following the Three Mile Island accident. Various measures had been implemented to enhance safety and restore confidence in the industry, and around 70 per cent of the TMI Action Plan was implemented, further improving safety protocols.[27] While the aftermath of Three Mile Island had led to a drop in orders, the overall trend showed a positive outlook for the industry. In the United States, as of June 1983, power reactors accounted for approximately 13 per cent of the country's total electricity generation. Globally, there were

277 reactors in operation. And industry reports from that time projected a significant increase in nuclear capacity, with more countries considering the adoption of nuclear energy for electricity generation.[28]

One of the countries who had an impressive nuclear energy programme was the Soviet Union. With the initiation of APS-1 they had been the first country in Europe to operate a reactor, and by the mid-1980s they had 43, with an additional 36 undergoing construction and 34 in the planning phase.[29] Among these was the Chernobyl Nuclear Power Plant, just over a hundred kilometres north of Kyiv. Then part of the Soviet Union, the plant comprised of four water-cooled, graphite-moderated reactors known as RBMKs. Units 1 and 2 had been constructed between 1970 and 1977, while units 3 and 4, sharing the same design, were completed in 1983. There were two more reactors also under construction.

On 26 April 1986, while undergoing tests, an incident occurred at Reactor 4. During this time, while the emergency safety systems had been intentionally turned off, a sudden power surge triggered a steam explosion and fire. The explosion resulted in the ignition of the graphite blocks shielding the uranium, causing a devastating blaze and the destruction of the reactor core building.[30]

Soviet officials restricted access to the nuclear plant and eventually managed to control the fire.[31] Firefighters were dispatched to extinguish the blazes on the roof, which were successfully put out within a few hours. However, it is worth noting that these firefighters were not given any sort of protection against radioactivity and were only wearing their regular attire. The initial response involved the power station's own firefighters, followed by additional assistance from nearly 200 firefighters who arrived from various locations in the region.

It was not until the following day that a ten-kilometre exclusion zone was established, and authorities initiated the evacuation of Prypiat, the city where the majority of the workers and their families lived. Residents were informed through loud-speakers about the incident and urged to leave, with instructions to pack essential belongings, food and clothes. Reassurances were given that it was a precautionary step, and they would be able to return soon. Subsequently, the exclusion zone was expanded to a 30-kilometre radius, leading to the evacuation of approximately 116,000 people.

The Soviet authorities had initially tried to conceal the accident not only from its citizens, but from the world.[32] However the incident came to light when engineers at a nuclear power station in Sweden detected elevated radiation levels on their instruments.[33] At the Forsmark Nuclear Power Plant, an employee arrived for work and triggered alarms as they passed through the radiation-monitoring portal. It was realised that his clothing was contaminated with radioactive dust.[34] Subsequently, Swedish authorities conducted an analysis linking elevated radiation levels across Europe with wind patterns. They publicly announced that a nuclear incident had taken place in an undisclosed location within the Soviet Union. And while the thought of undisclosed radioactivity is one of the things that terrifies people, it is important to remember that one of the truly amazing things about radiation is how easily and quickly even very small amounts can be detected. Unlike Three Mile Island, the reactor had only a partial containment building. Consequently, there was nothing to stop the release of a cloud of radioactive smoke, containing isotopes like the short-lived iodine-131 and long-lived caesium-137, which was released into the atmosphere. Carried by the winds, the lighter particles dispersed across Ukraine, Belarus, Russia, Scandinavia and, to a degree, throughout Europe.[35] They increased the level of radiation in the atmosphere in these affected places but quickly receded without any subsequent measurable increased cancer risk or health problems to their populations.

After three days the Soviet government, in a statement read over Radio Moscow, announced that 'an accident has occurred at the Chernobyl atomic power plant as one of the atomic reactors was damaged. Measures are being taken to eliminate the consequences of the accident. Aid is being given to those affected. A government commission has been set up.'[36]

The next priority was to clean up the radioactivity at the site. A workforce of 600,000 individuals, known as liquidators – which comes from the Russian word 'likvidator,' meaning 'to eliminate the consequences of an accident' – were sent from all over the Soviet Union to take part in what would become an initial three-year clean-up programme culminating in the construction of a massive sarcophagus made of concrete and steel to enclose the destroyed reactor. While the initial explosion had led to the tragic death of two people, a staggering 28 of the emergency clean-up workers and firemen were heavily irradiated and died in the first three months after the accident. The cause of these deaths was acute radiation sickness, an illness caused when people receive high doses of penetrating radiation over their entire bodies in a very short space of time. Depending on exposure it can be fatal, but it can be treatable and ultimately recoverable.

In the aftermath of Chernobyl, following the pattern seen after Three Mile Island, there were widespread protests and demonstrations against nuclear energy and weapons production. These protests took place in various countries, including Italy, West Germany and France, targeting nuclear energy plants and used-fuel reprocessing facilities. Additionally, the issue of nuclear disarmament gained significant attention during this time. In 1986, the Great Peace March for Global Nuclear

Disarmament was organised, where hundreds of people walked from Los Angeles to Washington, DC, covering a distance of 3,700 miles over a span of nine months.

The backlash against all forms of nuclear energy was also accompanied by calls for historical compensation. Communities affected by nuclear fallout or living near wartime nuclear facilities sought recognition and damages for the health risks they had experienced.

These campaigns were often the culmination of decades-long struggles to hold the government and industry accountable for its actions and to address social and environmental impacts. Furthermore, emerging in the early 1970s, ongoing concerns about the effects of uranium mining had sparked the formation of a network of Navajo activists, including former workers, family members and health professionals. This network eventually developed into the Navajo Office of Uranium Workers, advocating for redress and compensation for those affected.[37] A key piece of legislation was the passing of the Radiation Exposure Compensation Act (RECA), which was enacted by Congress in 1990 and was to provide compensation to those who had developed cancer or respiratory ailments due to radiation exposure. Initially, the act focused on miners who had been exposed to radiation while extracting uranium, as well as some downwinders, offering them 'compassion payments' of $100,000. However, the act was later amended to include mill workers and ore transporters as well, recognising their contribution to the war effort.[38] It was amended again in July 2023 to finally include families exposed to radioactivity by the Trinity test and workers during the Cold War era. It is still not enough.

While RECA was an acknowledgement of the troubled legacy of the history of uranium production for both defence and civilian use, there were still – and still are today – public health impacts and environmental consequences from this history that have not been fully addressed. Dealing with these issues is vital not only to right historic wrongs – there are estimated to be over

500 abandoned uranium mines on and near the Navajo Nation alone – but also for managing our growing demand for uranium, which is expected to rise in parallel with the need for low carbon energy in the future. With 70 per cent of global uranium deposits located on traditional lands of indigenous peoples in the uranium-producing countries of Australia, Kazakhstan, Russia, Canada, Niger and South Africa, implementing sustainable and ethical mining practices and policies has to be at the forefront of any serious discussion about the future of nuclear energy.[39]

While some countries, particularly France and Japan, continued to build in the late 1980s, many took nuclear energy off their policy agenda, and several halted or postponed the construction of new nuclear power plants.[40] Some countries pledged to phase out existing reactors and to stop building new ones. Following a series of referendums in 1987, Italy began shutting down all its nuclear power plants the following year. Despite Chernobyl, RBMK technology was far from dead. The International Atomic Energy Agency (IAEA) worked with countries that generated their energy using RBMK reactors and upgraded them, adding in new protection and correcting the known flaws of the technology. At Chernobyl the other three reactors continued to operate and the plant's performance actually improved, producing more electricity in 1988 than it had in 1986. The last reactor on site was eventually shut down in 2000.

On 11 March 2011, a powerful 9.0-magnitude earthquake struck the east coast of Japan. The Tōhoku earthquake, which was the strongest recorded in the country's history, was followed by a tsunami that reached up to 30 metres in some areas, sweeping entire towns away and killing nearly 20,000 people, including three employees at nuclear power plants.

At the time of the earthquake, Japan, which first began developing nuclear energy in the mid-1950s and launched its first commercial plant in 1965, had seventeen sites operating 55 reactors and produced about a third of all the country's electricity.[41]

Within the immediately affected area were three nuclear energy plants: Fukushima Daiichi, Fukushima Daini and the Onagawa nuclear power station. At Fukushima Daini and Onagawa, despite being close to the epicentre of the earthquake and experiencing a higher tsunami wave, operators were able to cold shutdown all operating reactors without incident – the installed safety features working as designed. But at Fukushima Daiichi there was a very different outcome. While executives from the Tokyo Electric Power Company (TEPCO) were eventually cleared of professional negligence, a series of design choices dating to the plant's construction in the 1970s coupled with a failure to implement more modern safety measures had left the site vulnerable. The effect of the earthquake followed by a series of tsunami waves triggered a chain of events that led to the fuel in the reactors overheating and a significant off-site release of radiation. In the days that followed, evacuation orders were issued and it was declared a 'major accident' on the IAEA International Nuclear and Radiological Event Scale (INES). It was not until December 2011 that Japanese Prime Minister Yoshihiko Noda declared that the cold shutdown of the reactors was complete, and he paid tribute to those who had bought the plant under control.

Mass demonstrations followed in places like Britain, China and Japan, calling for a complete end to nuclear energy. Several nuclear power plants in the United States were shut down, including San Onofre Units 2 and 3 in California, Crystal River Unit 3 in Florida, and Kewaunee in Wisconsin. Switzerland, a country who had held many referendums on the question since Three Mile Island, decided to freeze any new reactors and not to replace those that were currently operating. The last of their reactors is due to go offline in 2034.

Japan made the decision to close 54 of its reactors, significantly reducing its nuclear energy generation to around 1 per cent, down from a previous high of 30 per cent of its total. Germany almost immediately announced it was completely phasing out nuclear power. While this was a dramatic decision, the country had been struggling with its relationship with nuclear energy for many years. Despite the fact that around a fifth of the country's electricity was generated by nuclear, they had, in 2002, announced that they intended to phase it out by 2022. In 2010, Chancellor Merkel had announced that this would be delayed until 2036, a decision reversed in the light of Fukushima. This staggered closing, which finally took full effect in April 2023, was part of a wider policy known as *Energiewende* – a transition from fossil fuels and nuclear power to renewables.

As such, sources of energy from renewable resources, like solar, hydro and geothermal heat, received huge subsidies and by 2020 made up more than 50 per cent of Germany's energy mix. The remainder is still powered by fossil fuels and the phasing out of Germany's coal-fired power stations is not scheduled until 2038.

And this is at the heart of the issue, because nuclear power is almost always replaced by fossil fuels, and results in an increased reliance on this to generate electricity. In Germany's case it has led to the country becoming the European Union's largest consumer of coal-generated electricity. And, despite significant investments in renewables, they have not reached their climate targets.

The health impacts of carbon dioxide as well as the other emissions like mercury and uranium produced by burning fossil fuels are well documented. At their most minor they can cause asthma, nausea and respiratory illnesses, but can often prove fatal. And this is important because ash from coal power plants contains uranium and thorium, carrying radiation into the environment at a much larger rate than any from a nuclear power plant. An analysis conducted in 2019, examining the post-Fukushima energy policies of Germany and Japan, showed that both countries could have prevented approximately 28,000 deaths caused by air pollution and reduced the emission of 2,400 metric tons of carbon dioxide if they had retained their existing nuclear power capacity instead of relying on fossil fuels to compensate for the nuclear shutdowns.[42]

Fossil fuels not only kill millions of people each year, but have a terrible safety record when it comes to extraction and processing. Coal mining destroys habitats, trees are cut or burned down, topsoil is scraped away leading to soil erosion. It exposes the miners and the environment to toxins like mercury, arsenic and many other nasties. Inhaled coal dust causes lung disease. And, of course, there are the inherent dangers in mining, leading to accidents and death. This can mean horrific outcomes for the individual miners, but also for whole towns. In 1966 the Aberfan colliery disaster in Wales killed almost 150 people including 116 children, and only a few years later, in West Virginia, there was the Buffalo Creek flood where a 30-foot-high tsunami of coal waste hit sixteen towns, killing 125 people.[43]

In a strange way it is the rarity of nuclear incidents that make them notable – so the names Chernobyl, Fukushima and (to a lesser extent) Three Mile Island are burned in our minds, used as shorthand for disasters. Nuclear power accidents are scary and nothing if not memorable – even to the point that we might not always recall them accurately. After all, Fukushima is largely remembered as a nuclear disaster rather than the natural disaster – the tsunami and earthquake – which caused tens of thousands of deaths and massive environmental destruction. This reaction is why we remember Chernobyl with horror but there is not the same attitude towards the chemical accident at the Union Carbide pesticide plant in Bhopal, India, two years earlier, which exposed half a million people to toxic gas and killed an estimated 25,000 people.[44] Or the Deepwater Horizon oil spill in 2010, or the 2023 Ohio train derailment, which released hydrogen chloride and phosgene into the air with disastrous health consequences.

It's why when Russian forces destroyed the Kakhovka dam in Ukraine in June 2023, the focus of reporting was on the potential of what may happen to the nearby Zaporizhzhia nuclear power station and not the thousands of people forced from their homes, the missing and the dead. It certainly didn't lead to protests demanding the end of hydropower – which in itself has both extreme environmental impacts while being an incredibly beneficial and largely safe means of electricity generation. We didn't see the same mass protests after the collapse of the Banqiao Dam in China in 1975 – one of the most catastrophic energy-generation disasters of all time. This event resulted in the tragic loss of between 171,000 and 230,000 lives, yet this is rarely brought up when hydropower is championed.

It's why we laugh along with the continual disasters portrayed at the Springfield Nuclear Power Plant (SNNP), which in different episodes of the long-running animated sitcom *The Simpsons* commits numerous safety violations, including leaking pipes, radioactive waste, fuel rods used as paperweights, and the creation of a mutant subspecies of three-eyed fish known as 'Blinky', when in reality none of these things can happen. Not even the operator accidentally pressing a self-destruct button while asleep! Even the Playmates SNNP toy, which was released in 2000, has a button that triggers one of 40 trademarked phases, including: 'Run for your lives everyone, this is not a drill.'

It's important to remember that, despite those high-profile incidents, nuclear's safety record over the last seven decades (18,000 reactor years) is comparable with solar and wind. In fact, nuclear is one of the safest forms of electricity produced – only beaten by solar.

Ultimately, we understand that things go wrong – that systems fail, that improvements need to be made. It's why we still get on planes despite how many losses and accidents there were when the technology was first developed, and how it still can be subject to catastrophic failures and human error, not to mention terrorist attacks. We are incredibly bad at working out the real risks and separating out what has gone before and what can happen in the future – especially, so it appears, when it comes to nuclear energy.

And when you consider the language we use, this isn't surprising – we just can't get away from atomic bombs and destruction. So, we use the word 'nukes' as a shorthand for both weaponry and power plants. Political parties and campaigning groups run on platforms of both the excellent and laudable aim of nuclear non-proliferation and ending commercial nuclear power, without really justifying why they should still be so inextricably linked. That there are nine nuclear armed states which between them possess more than 12,500 nuclear warheads is a huge world problem.

And it's no wonder when you consider that the linear no-threshold hypothesis still informs so many of the assumptions, fears and regulations surrounding radiation. The adherence to the As Low As Reasonably Achievable (ALARA) principle is seen by many as harmful because it perpetuates the public perception that even low levels of radiation are highly dangerous. This, in turn, leads to extreme precautionary measures being implemented.

We ignore evidence that doesn't fit in with the narrative and allow fear to make extreme decisions. Fukushima led to 150,000 people evacuating their homes. Figures vary, but it is estimated that up to 1,600 people died due to the hasty, poorly managed and stressful evacuation.[45] The psychological effects on the surviving evacuees should also not be underestimated – many of them have not returned to their homes, despite the fact that only inconsequential amounts of radiation had been released into the environment. Fishing near the plant was banned, farmers were told to euthanise their cattle and not to grow rice.

Such was the fear of radioactive contamination that many countries, including China and South Korea, implemented bans on products from the region. Even a decade later the panicked response is still causing economic harm, with producers reporting an immediate drop off in sales coupled with a long-term resistance from consumers concerned about buying anything made or grown in the Fukushima Prefecture, an area of 13,000 square kilometres. It was only in June 2022 that Britain lifted import restrictions on products, including the famous saké from the region, there are still limitations in place in other countries.

And while Chernobyl was also subject to an evacuation order 40 years ago, it isn't quite the desolate, uninhabitable place that nuclear energy opponents would have it. While the Zone of Alienation (exclusion zone) is largely off limits, there are a number of 'self-settlers', people who returned to the zone and live a largely self-sufficient life. And no, they don't glow in the dark.

There are also the thousands of people who work in the community, from security guards to scientists, as well as the tens of thousands of tourists who visit the site each year, or at least used to before the most recent Soviet invasion of Ukraine. These nuclear tourists, armed with permits and guides, see the site as well as buying souvenirs and merchandise, including 'radioactive' ice cream and cans of Chernobyl air, which allow you to inhale 'the unforgettable smell of abandoned concrete structures of the Soviet Union, the dampness of basements, mixed with the aroma of Pripyat roses'. While Chernobyl is currently off limits it is still possible to support those who live and work there by buying a bottle of Atomik, a spirit made with fruit grown in the area.[46]

But it is what has happened to the wildlife that is the most fascinating aspect – the roughly 2,800 square-kilometre area of northern Ukraine is now the third-largest nature reserve in mainland Europe, with thriving populations of wild boar, elk, wolves and rare wild Przewalski's horses. It's a testament not to the destructive potential of nuclear energy, but to the benefits of removing humans generally.

Our long memory for nuclear accidents and over-exaggerated fears of radioactivity are even more remarkable when you consider how much radiation we encounter every day of our lives: from X-rays to flights to the background levels of radiation where you live.

For many people their exposure to natural radiation is greater than from any other source. Guarapari in Brazil is home to naturally radioactive beaches, courtesy of eroded Monazite, an ore containing thorium. It's a popular tourist destination which sells itself not only for the gorgeous coastline, but for its supposed therapeutic properties. In fact, my Brazilian-born husband can remember his grandmother going there and covering herself with the sand. It is definitely on my holiday bucket list as well.

And while the LNT model tells us that even the smallest dose of radiation can be harmful, not all experts agree with this. There is no evidence to show that any of these naturally radioactive areas are associated with elevated levels of cancer or tumours. Some researchers even suggest that low levels of radiation can be helpful to human health, with experiments on how it may even be critical for normal cell function and genomic stability. Radon spas and radon therapy mines are still in existence: not just the healing tunnel in Bad Gastein, but also the Radium Palace, Czech Republic and the Free Enterprise Health Mine in Boulder.

But, even if you are not swayed by such arguments, it is remarkable that any extra radiation above what constitutes 'background' levels is met with protest and criticism – disproportionate to the risks involved. As I write this, there remains a huge backlash around planned releases of water from the Fukushima Daiichi plant. This water was used to cool the melted fuel and debris in 2011 and has been safely stored in tanks on site ever since. It has also been cleaned by a technology called Advanced Liquid Processing System (ALPS), which removes contaminates, including radioactive iodine and caesium. ALPS is hugely effective technology but still leaves trace amounts of tritium, a radioactive form of hydrogen that is so chemically similar to water it is difficult to remove. But that doesn't mean it is dangerous – not only is tritium already found in seawater, the water released is so diluted that it is well below safe levels and has no impact on the Pacific Ocean or its contents. I can't help but think if only there was as much emphasis on mercury and arsenic contamination, which is found in seawater as a result of less well-regulated industries.

Despite these occasional outcries, today there are many powerful advocates for nuclear energy. These include Isabelle Boemeke, the world's first nuclear-energy influencer, Grace Stanke, a nuclear engineering student and Miss America 2023, Stand up for Nuclear, and Mothers for Nuclear to name a few. Even the Green Party in the UK has its own subsect of members, Greens for Nuclear Energy, dedicated to amending one of the party's central policies – EN014: 'Nuclear power, coal, and incineration of waste will be phased out'. Filmmakers Oliver Stone and Frankie Fenton have also put together the powerful documentaries *Nuclear Now* and *Atomic Hope: Inside the Pro Nuclear Movement* respectively, which were released to great acclaim. Environmental campaigners like George Monbiot and Stewart Brand have revised their past stance on nuclear energy and its role in mitigating climate change. Polls have shown an increase in support for new nuclear power in the US, France and the UK to name a few. Even in Germany, surveys showed that over 70 per cent of the public supported the continued use of nuclear energy. These results were taken only a few months before their final three plants were shut down. The good news is that many of Germany's closed reactors are in excellent condition, as required by German environmental law, and can be rapidly brought back into use should the political mood alter – as long as they are not demolished, of course!

As we strive to combat climate change and embrace low-carbon energy sources, the ability to split atoms and provide a reliable baseload power supply will be crucial for meeting continuous electricity demand and maintaining grid stability. This is especially significant as other energy sources, such as wind and solar, will always require a complementary and stable energy source to ensure consistent power supply.

In combination with nuclear power, we can build a comprehensive energy mix that can effectively address the global energy transition towards a cleaner and more sustainable future. By integrating nuclear energy alongside solar, wind and other renewable sources, countries can achieve a diversified and resilient energy portfolio that supports their climate objectives. And, crucially, move away from fossil fuels at a quicker rate.

Many countries, including China, Turkey, France, India, Canada, the United Arab Emirates and the United Kingdom are investing heavily in nuclear power to achieve their climate goals and to address increasing energy demands. Japan is restarting their reactors again – albeit slowly. In May 2023 their parliament enacted a bill to allow nuclear reactors to be operated beyond the current limit of 60 years. A much-delayed reprocessing plant is also due to begin commercial operation in 2024. As of 2024, there are over 440 commercial nuclear power reactors spanning 32 countries worldwide, with lots more currently under construction or on order.[47] These reactors play a crucial role in generating approximately 10 per cent of the world's electricity.[48] They can also be used to power desalination plants and even generate clean hydrogen as a fuel for vehicles, replacing hydrogen from natural gas and its resultant carbon emissions. Additionally, research reactors serve important purposes in the production of medical and industrial isotopes.

Expansion of nuclear energy faces hurdles such as public acceptance, regulatory challenges and competition from renewable energy sources. In many countries there are a shortage of skilled workers, and the supply of nuclear fuel have emerged as a significant concern.

As befitting a still emerging technology, there are some hugely exciting innovations in development, including the long-promised fusion power, advanced reactor designs cooled by molten salts, high temperature gases or liquid metal – not to mention modular reactors. These reactors are designed to be flexible, cost effective, scalable and can be built in factories and then assembled on site, reducing cost and development time. They are also powerful energy generators that do not require too much land – something that other renewable power sources struggle with in comparison – and can be adapted to civilian maritime use, helping to decarbonise the highly polluting shipping industry. And maybe, one day, we may even see a revival of the concept highlighted by the NS *Savannah* and see cruise ships powered by atom splitting.

But whether we are seeing the beginnings of a renaissance – a second hopeful history – and whether the nuclear industry can truly innovate, change hearts and minds in time to address our present climate emergency, is a question for a historian to tackle twenty years from now.

But in the meantime, as the updated protest badge says: 'Nuclear Power? Yes please.'

ACKNOWLEDGEMENTS

With the working title #UraniumBook, this project was envisioned during late 2020, a time of confusion, sacrifice, misinformation and a little bit of hope for a kinder, more compassionate world.

For me it was four months after my debut book, *Half Lives: The Unlikely History of Radium,* had been published and arrived in a world where launch parties were virtual and book shops were closed.

The team at Icon Books were magnificent during this period so a huge thanks to Ruth Killick, Hamza Jahanzeb, Duncan Heath, Robert Sharman and Andrew Furlow. I was delighted when they agreed to take me on for this, my second book, and for the care and attention given by Connor Stait and Steve Burdett.

Thanks to Laura Macdougall and Olivia Davies at United Agents who provided encouragement and wise advice every step of the way. And my media agent Nigel Hetherington of Past Preservers and Corey William Schneider of the New York Adventure Club, who have found me some wonderful opportunities to talk about many of the topics covered here.

To librarians, archivists, curators and volunteers at the National Atomic Testing Museum; National Archives, Kew; British Library; Wellcome Collection; the Centre for Indigenous and Settler Colonial Studies, University of Kent American Medical Association. And to my beta readers who generously took the time to read and comment.

And to all my family and friends, especially Mike, Sarah, Jane, Penny, Shareen, Mark, Sasha, Liam, Ryan, Katrina, Andrew, Lily and Freya. Alice, Krishna, Kate, Anne, Charlie and James, Emma and Billy, Laura, Hillary, Natalie, Fiona and Michelle. And, of course, especially, to Amber.

Finally, to Andy, my husband and biggest supporter. Sorry for all the weird places I make you go to, but at least we got to have a spa at the same place as Arnie.

BIBLIOGRAPHY

Len **Ackland**. *Making a Real Killing: Rocky Flats and the Nuclear West*, University of New Mexico Press, 1999.

Amir D. **Aczel**. *Uranium Wars: The Scientific Rivalry that Created the Nuclear Age*, Palgrave Macmillan, 2009.

Jon **Agar**. *Science in the 20th Century and Beyond*, Polity Press, 2012.

Georgius **Agricola**. *De Re Metallica: Translated from the First Latin edition of 1556 by Herbert Clark Hoover and Lou Henry Hoover*, Dover Publications, 1950.

Becky **Alexis-Martin**. *Disarming Doomsday: The Human Impact of Nuclear Weapons since Hiroshima*, Pluto Press, 2019.

Michael A. **Amundson**. *Yellowcake Towns: Uranium Mining Communities in the American West*, University of Colorado, 2011.

Herbert L. **Anderson**. *John Ray Dunning: A Biographical Memoir*, National Academy of Sciences, 1989.

Peter **Bacon Hales**. *Atomic Spaces: Living on the Manhattan Project*, University of Illinois Press, 1997.

Peter **Bacon Hales**. *Outside the Gates of Eden: The Dream of America from Hiroshima to Now*, University of Chicago Press, 2014.

Lawrence **Badash**. *Radioactivity in America: Growth and Decay of a Science*, Johns Hopkins University Press, 1979.

Lennard **Bickle**. *The Deadly Element: The Story of Uranium*, Macmillan, 1980.

Joanna **Bourke**. *Fear: A Cultural History*, Virago, 2015.

Paul S. **Boyer**. *By the Bomb's Early Light: American Thought and Culture at the Dawn of the Atomic Age*, Pantheon, 1985.

Deborah **Blum**. *Ghost Hunters*, Penguin Press, 2007.

Ruth Brandon. *The Burning Question: The Anti-nuclear Movement Since 1945*, William Heinemann Ltd, 1987.

Ruth Brecher and Edward M. Brecher. *The Rays: History of Radiology in the United States and Canada*, Lippincott Williams & Wilkins, 1969.

William H. Brock. *William Crookes (1832–1919) and the Commercialization of Science*, Ashgate, 2008.

Peter Broks. *Media Science Before the Great War*, Macmillan Press, 1996.

G.I. Brown. *Invisible Rays: A History of Radioactivity*, Sutton Publishing Ltd, 2002.

Jerry Brown and Rinaldo Brutoco. *Profiles in Power: The Antinuclear Movement and the Dawn of the Solar Age*, Twayne Publishers, 1997.

Traci Brynne Voyles. *Wastelanding: Legacies of Uranium Mining in Navajo Country*, University of Minnesota Press, 2015.

Alice Buck. *The Atomic Energy Commission*, AEC, 1983.

Michael Burgan. *Chernobyl Explosion: How a Deadly Nuclear Accident Frightened the World*, Capstone Press, 2018.

Anthony Burke. *Uranium*, Polity Press, 2017.

John Campbell. *Rutherford: Scientist Supreme*, AAS Publications, 1999.

Catherine Caufield. *Multiple Exposures: Chronicles of the Radiation Age*, Secker & Warburg, 1989.

Kit Chapman. *Superheavy: Making and Breaking the Periodic Table*, Bloomsbury Sigma, 2021.

Claudia Clark. *Radium Girls: Women and Industrial Health Reform, 1910–1935*, University of North Carolina Press, 1997.

Gwyneth Cravens. *Power to Save the World: The Truth About Nuclear Energy*, Alfred A Knopf, 2007.

Angela N.H. Creager. *Life Atomic: A History of Radioisotopes in Science and Medicine*, University of Chicago Press, 2013.

Hunter Davies. *Sellafield Stories*, Constable, 2012.

Aymon De Lestrange. *Coca Wine: Angelo Mariani's Miraculous Elixir and the Birth of Modern Advertising*, Park Street Press, 2018.

Finis Dunaway. *Seeing Green – The Use and Abuse of American Environmental Images*, University of Chicago Press, 2015.

Peter H. Eichstaedt. *If You Poison Us: Uranium and Native Americans*, Red Crane Books, 1994.

Maxwell Leigh Eidinoff. *Atomics for the Millions*, McGraw-Hill, 1947.

Eduard Farber. *Great Chemists*, John Wiley & Sons, 1962.

Laura Fermi. *Atoms for the World: United States Participation in the Conference on the Peaceful Uses of Atomic Energy*, University of Chicago Press, 1957.

The Focal Encyclopedia of Photography, The Macmillan Co, 1960.

Paul Frame & William Kolb. *Living With Radiation: The First Hundred Years*, Syntec, Inc, 2002.

Sir James George Frazer, trans. *Ovid's Fasti*, William Heinemann Ltd, 1959.

Lindsey A. Freeman. *Longing for the Bomb: Oak Ridge and Atomic Nostalgia*, University of North Carolina Press, 2015.

Conrad Gessner. *De omni rerum fossilium genere, gemmis, lapidibus metallis, et huiusmedi libri aliquet, plerique nunc primum editi*, Tiguri, 1565.

C.S. Gilbert. *An Historical Survey of the County of Cornwall Vol 1*, J Congdon: Plymouth, 1817.

Marco Giugni. *Social Protest and Policy Change: Ecology, Antinuclear, and Peace Movements in Comparative Perspective*, Rowman & Littlefield, 2004.

Joshua S. Goldstein and Staffan A. Qvist. *A Bright Future: How Some Countries Have Solved Climate Change and the Rest Can Follow*, Hachette Book Group, 2020.

Margaret Gowing. *How Nuclear Power Began*, University of Southampton, 1987.

Larry D. Gragg. *Becoming America's Playground: Las Vegas in the 1950s*, University of Oklahoma Press, 2019.

Peter Harclerode. *Warfare*, Channel 4 Books, 2000.

James Harvey Young. *The Toadstool Millionaires*, Princeton University Press, 1972.

David I. Harvie. *Deadly Sunshine: The History and Fatal Legacy of Radium*, Tempus, 2005.

Gabrielle Hecht. *Being Nuclear: Africans and the Global Uranium Trade*, MIT Press, 2014.

Eva Hemmungs Wirtén. *Marie Curie: Intellectual Property and Celebrity Culture in an Age of Information*, University of Chicago Press, 2015.

Jonathan Hogg. *British Nuclear Culture: Official and Unofficial Narratives in the Long 20th Century*, Bloomsbury Academic, 2016.

Joel D. Howell. *Technology in the Hospital: Transforming Patient Care in the Early Twentieth Century*, Johns Hopkins University Press, 1995.

Jeff A. Hughes. *The Manhattan Project: Big Science and the Atom Bomb*, Icon Books, 2003.

Maxwell Irvine. *Nuclear Power: A Very Short Introduction*, Oxford University Press, 2011.

E.N Jenkins. *Radioactivity: A Science in Its Historical and Social Context*, Butterworth & Co, 1979.

Robert Johnson. *Romancing the Atom: Nuclear Infatuation from the Radium Girls to Fukushima*, Praeger, 2012.

Barbara Rose Johnston. *Half-lives and Half-truths: Confronting the Radioactive Legacies of the Cold War*, School for Advanced Research Press, 2007.

Sean F. Johnston. *History of Science: A Beginner's Guide*, Oneworld Publications, 2009.

Christian Joppke. *Mobilizing Against Nuclear Energy: A Comparison of Germany and the United States*, University of California Press, 1993.

Timothy J. **Jorgensen**. *Strange Glow: The Story of Radiation*, Princeton University Press, 2016.

Denise **Kiernan**. *The Girls of Atomic City: The Untold Story of the Women Who Helped Win World War II*, Atria Books, 2013.

Harry **Kursh**. *How to Prospect for Uranium*, Fawcett Publications, 1955.

William **Lanouette** with Bela Silard. *Genius in the Shadows: A Biography of Leo Szilard, the Man behind the Bomb*, University of Chicago Press, 1994.

J.L. **Lewis** and E.J. Wenham. *Radioactivity*, Longman, 1970.

Ruth **Lewis Sime**. *Lise Meitner: A Life in Physics*, University of California Press, 1996.

Al **Look**. *U-boom: Uranium on the Colorado Plateau*, Bell Press, 1956.

James A. **Mahaffey**. *Atomic Awakening: A New Look at the History and Future of Nuclear Power*, Pegasus Books, 2009.

Marjorie C. **Malley**. *Radioactivity: A History of a Mysterious Science*, Oxford University Press, 2011.

Leona **Marshall Libby**. *The Uranium People*, Russak, 1979.

Bob **McCoy**. *Quack!: Tales of Medical Fraud from the Museum of Questionable Medical Devices*, Santa Monica Press, 2000.

Richard F. **Mould**. *A History of X-Rays and Radium with a Chapter on Radiation Units 1895–1937*, IPC Building & Contract Journals, 1980.

Richard F. **Mould**. *A Century of X-Rays and Radioactivity in Medicine: With Emphasis on Photographic Records of the Early Years*, CRC Press, 1993.

Craig **Nelson**. *The Age of Radiance: The Epic Rise and Dramatic Fall of the Atomic Era*, Scribner Book Company, 2014.

Andrew **Norman**. *The Amazing Story of Lise Meitner: Escaping the Nazis and Becoming the World's Greatest Physicist*, Pen and Sword, 2021.

Daniel **Novak**. *Realism, Photography and Nineteenth Century Fiction*, Cambridge University Press, 2008.

James S. **Olson**. *Bathsheba's Breast: Women, Cancer and History*, Johns Hopkins University Press, 2002.

John **O'Neill**. *Almighty Atom: The Real Story of Atomic Energy*, Ives Washburn Inc, 1945.

M. Alice **Ottoboni**. *The Dose Makes the Poison: A Plain Language Guide to Toxicology*, Van Nostrand Reinhold, 1991.

Judy **Pasternak**. *Yellow Dirt: An American Story of a Poisoned Land and a People Betrayed*, Free Press, 2010.

Fred **Pearce**. *Fallout: Disasters, Lies and the Legacy of the Nuclear Age*, Beacon Press, 2018.

Diana **Preston**. *Before the Fall-Out: The Human Chain Reaction from Marie Curie to Hiroshima*, Doubleday, 2005.

Jerome **Price**. *The Antinuclear Movement*, Twayne Publishers, 1982.

Diana Řeháčková. *Jáchymovské radiové lázně v minulosti a přítomnosti*, Diana Řeháčková, 2013.

Richard **Rhodes**. *Energy: A Human History*, Simon & Schuster, 2018.

Iwan **Rhys Morus**. *Shocking Bodies: Life, Death & Electricity in Victorian England*, The History Press, 2011.

Raye C. **Ringholz**. *Uranium Frenzy: Boom and Bust on the Colorado Plateau*, Norton, 1989.

Teresa **Riordan**. *Inventing Beauty: A History of the Innovations that Have Made Us Beautiful*, Broadway Books, 2004.

Roger F. **Robison**. *Mining and Selling Radium and Uranium*, Springer, 2015.

Kenneth D. **Rose**. *One Nation Underground: The Fallout Shelter in American Culture*, New York University Press, 2001.

Lucy Jane **Santos**. *Half Lives: The Unlikely History of Radium*, Icon Books, 2020.

Secret Remedies: What They Cost and What They Contain, British Medical Journal, 1909.

Robert **Sharp**. *Early Days in Katanga*, Rhodesian Printers, 1956.

Tim **Smolko** and Joanna Smolko. *Atomic Tunes: The Cold War in American and British Popular Music*, Indiana University Press, 2021.

Raymond W **Taylor**. *Uranium Fever: or, No Talk Under $1 Million*, Macmillan, 1970.

A. Costandina **Titus**. *Bombs in the Backyard: Atomic Testing and American Politics*, University of Nevada Press, 2001.

Claudio **Tuniz**. *Radioactivity: A Very Short Introduction*, Oxford University Press, 2012.

J Samuel **Walker**. *Containing the Atom: Nuclear Regulation in a Changing Environment, 1963–1971*, University of California Press, 1992.

Spencer R. **Weart** and Gertrud Weiss Szilard (eds.). *Leo Szilard, His Version of the Facts: Selected Recollections and Correspondence*, MIT Press, 1978.

Alvin Martin **Weinberg**. *The Second Nuclear Era: A New Start for Nuclear Power*, Praeger, 1985.

Thomas Raymond **Wellock**. *Critical Masses: Opposition to Nuclear Power in California 1958–1978*, The University of Wisconsin Press, 1998.

Gerald **Wendt**. *The Atomic Age Opens*, The World Publishing Company, 1945.

David **Wilson**. *Rutherford: Simple Genius*, MIT Press, 2004.

Allan M. **Winkler**. *Life under a Cloud: American Anxiety about the Atom*, Oxford University Press, 1993.

Jeffrey **Womack**. *Radiation Evangelists: Technology, Therapy and Uncertainty at the Turn of the Century*, University of Pittsburgh Press, 2020.

John **Woodforde**. *The Strange Story of False Teeth*, Routledge & Kegan Paul, 1983.

Scott C. **Zeman** & Michael A. Amundson. *Atomic Culture: How We Learned to Stop Worrying and Love the Bomb*, University Press of Colorado, 2004.

Tom **Zoellner**. *Uranium: War, Energy, and the Rock that Shaped the World*, Viking, 2009.

ARTICLES AND JOURNALS

L. **Badash** (1978) 'Radium, Radioactivity, and the Popularity of Scientific Discovery,' *Proceedings of the American Philosophical Society* (Vol 122 No 3), pp. 145–54.

Alison **Boyle** (2019) '"Banishing the Atom Pile Bogy": Exhibiting Britain's First Nuclear Reactor,' *Centaurus* (Vol 61, Issue 1–2), pp. 14–32.

George C. **de Hevesy** (1996) 'Marie Curie and Her Contemporaries,' *Journal of Nuclear Medical Technology* (Vol 24), pp. 273–79.

B. **Goldschmidt** (1962) 'France's Contribution to the Discovery of the Chain Reaction,' *Bulletin of the IAEA* (Vol 4–0), pp. 21–24.

R.T. **Günther** and J.J. Manley (1912) 'A Mural Glass Mosaic from the Imperial Roman Villa near Naples,' *Archaeologica* (Vol 63), pp. 99–108.

Jonathan E. **Helmreich** (1990) 'Belgium, Britain, The United States and Uranium 1952–1959,' *Studia Diplomatica* (Vol 43 No 3), pp. 27–81.

D. **Hogarth** (2014) 'Robert Richard Sharp 1881–1960: Discoverer of the Shinkolobwe Radium-Uranium Ore Bodies,' *The Journal of the Society for the History of Discoveries* (Vol 46, Issue 1), pp. 30–41.

E. **Joliot** and I. Curie (1934) 'Artificial Production of a New Kind of Radio-Element,' *Nature* (133), pp. 201–02.

Jerzy **Kunicki-Goldfinger** (2018) 'Uran w szkle historycznym,' *Szkło i Ceramika* (Vol 11), pp. 16–21.

Edward R. **Landa** (1987) 'Buried Treasure to Buried Waste: The Rise and Fall of the Radium Industry,' *Colorado School of Mines Quarterly*, pp. 34–63.

F. Peter **Lole** (1995) 'Uranium Glass in 1817: A Pre-Riedel Record,' *Journal of Glass Studies* (Vol 37), pp. 139–40.

L. **Meitner** and O. Frisch (1939) 'Disintegration of Uranium by Neutrons: A New Type of Nuclear Reaction,' *Nature* (Vol 143), pp. 239–40.

J.R. **Morgan** (1984) 'A History of Pitchblende,' *Atom* (No 329 March), pp. 63–68.

M.F. **Nelson** (1959) 'Use of Radioisotopes in Detergent and Cosmetic Research,' *Journal of the Society of Cosmetic Chemists*, November, pp. 320–30.

Dennis **Pohl** (2021) 'Uranium Exposed at Expo 58: The Colonial Agenda Behind the Peaceful Atom,' *History and Technology* (Vol 37), pp. 172–202.

M. **Rentetzi** (2008) 'The U.S Radium Industry: Industrial In-house Research and the Commercialization of Science,' *Minerva* (Vol 46), pp. 437–62.

Maria **Rentetzi** (2011) 'Packaging Radium, Selling Science: Boxes, Bottles and Other Mundane Things in the World of Science,' *Annals of Science* (Vol 68 No 3), pp. 375–99.

Roger F. **Robison** and Richard F. **Mould** (2006) 'St Joachimsthal: Pitchblende, Uranium and Radon Induced Lung Cancer,' *Journal of Oncology* (Vol 56 No 3), pp. 275–81.

J. Pohl **Rüling** (1993) 'The Scientific Development of a Former Gold Mine Near Badgastein, Austria to the Therapeutic Facility "Thermal Gallery",' in *Environment International: A History of Radium, Uranium, Thorium, and Related Nuclides in Industry and Medicine* (Vol 19 No 5), pp. 455–65.

E. **Rutherford** (1911) 'The Scattering of α and β Particles by Matter and the Structure of the Atom,' *Philosophical Magazine* (Series 6 Vol 21), May, pp. 669–88.

Frederick **Soddy** and Ruth **Pirret** (1910) 'The Ratio Between Uranium and Radium in Minerals,' *Philosophical Magazine* (Series 6 Vol 20), pp. 345–49.

Donna **Strahan** (2001) 'Uranium in Glass,' *Studies in Conservation* (Vol 46), pp. 181–85.

A. Costandina **Titus** (1983) 'A-Bombs in the Backyard: Southern Nevada Adapts to the Nuclear Age, 1951–63,' *Nevada Historical Society Quarterly* 26, p. 236.

Spencer **Weart** (1982) 'The Road to Los Alamos,' *Journal de Physique Colloques*, pp. 301–20.

EDITED COLLECTIONS

M. Rayner-**Canham** and G. Rayner-Canham (1997) 'And Some Other Women of the British Group,' in Marelene F. Rayner-Canham and Geoffrey W. Rayner-Canham (eds.) *A Devotion to their Science: Pioneer Women of Radioactivity*, McGill-Queen's University Press, pp. 156–61.

Jerry M. **Cuttler** (2015) 'Nuclear Energy and the LNT Hypothesis of Radiation Carcinogenesis,' in Shizuyo Sutou (ed) *Fukushima Nuclear Accident*, Nova Science Publishers, pp. 27–60.

Barbara **Erickson** (2000) 'And the People Went to the Caves to be Healed,' in Susanne Bentley, Brad Lucas and Stephen Tchudi (eds.) *Western Futures: Perspectives on the Humanities at the Millennium*, Nevada Humanities Committee, pp. 31–55.

F. **Habashi** (1997) 'Ida Noddack: Proposer of Nuclear Fission,' in Marelene F. Rayner-Canham and Geoffrey W. Rayner-Canham (eds.) *A Devotion to their Science: Pioneer Women of Radioactivity*, McGill-Queen's University Press, pp. 217–26.

Harold C. **Hodge** (1973) 'A History of Uranium Poisoning,' in H.C. Hodge, J.N. Stannard and J.B. Hursh (eds.) *Uranium-plutonium Transplutonic Elements*, Springer, pp. 5–68.

J. **Hudson** (2019) 'Dr Margaret Todd and the Introduction of the term 'Isotope,' in A. Lykknes and B. Van Tiggelen (eds.) *Women in their Element: Selected Women's Contributions to the Periodic System*, World Scientific Publishing, pp. 280–90.

L.P. **Jacquemond** (2019) 'Irène Joliot-Curie and the Discovery of "Artificial Radioactivity",' in A. Lykknes and B. Van Tiggelen (eds.) *Women in their Element:*

Selected Women's Contributions to the Periodic System, World Scientific Publishing, pp. 361–74.

Miloš René (2018) 'History of Uranium Mining in Central Europe,' in Nasser S. Awwad (ed) *Uranium – Safety, Resources, Separation and Thermodynamic Calculation*, IntechOpen, pp. 1–20.

Iwan Rhys Morus (2016) 'Making the Most Beautiful Experiment: Reconstructing Gassiot's Cascade,' in Martin Wills (ed) *Staging Science: Scientific Performance on Street, Stage and Screen*, Palgrave, pp. 11–35.

Lucy Jane Santos (2024) '"The Triumph of the New Over the Old": Electric Restaurants, Health and Modernity,' in Lauren Alex O'Hagan and Göran Eriksson (eds.) *Food Marketing and Selling Healthy Lifestyles with Science: Transhistorical Perspectives*. Forthcoming.

S. Watkins (1997) 'Lise Meitner: The Foiled Nobelist,' in Marelene F. Rayner-Canham and Geoffrey W Rayner-Canham (eds.) *A Devotion to their Science: Pioneer Women of Radioactivity*, McGill-Queen's University Press, pp. 163–92.

Tiffany Webber-Hanchett (2010) 'Bikini,' in Valerie Steele (ed) *The Berg Companion to Fashion*, Berg, pp. 77–79.

PHD/THESIS

Madeline Auerbach (2017) 'Dancing into the Mushroom Cloud: The Idealization of the Atomic Bomb in 1950s Las Vegas', BA Thesis, Georgetown University.

Edward E. Baldwin (1996) 'Las Vegas in Popular Culture', PhD Thesis, University of Nevada, Las Vegas.

David G. Präkel (2010) 'Not Fade Away – Lt. Col. Stuart Wortley, the United Association of Photography and the Wothlytype 1863–1867', Master's Thesis, De Montfort University.

Michael Pritchard (2010) 'The Development and Growth of British Photographic Manufacturing and Retailing 1839–1914', PhD Thesis, De Montfort University.

Joshua Samuel Scott Cornett (2017) 'Bombs, Bikinis and Godzilla: America's Fear and Fascination of the Atomic Bombs as Evidenced Through Popular Media', 1946–1962, Master's Thesis, Eastern Kentucky University.

Robert Wendell Holmes III (2010) 'Substance of the Sun: The Culture History of Radium Medicines in America', PhD Thesis, University of Texas at Austin.

ENDNOTES

Prologue
1. 4.6 billion years ago.

Chapter One
1. On a side note, his paper 'A Short, useful and very comforting regiment, how one can protect oneself with the help of God from the dwindling and poisonous plague of the pestilentz' sounds like it could be due a reprint.
2. Diana Řeháčková (2013) *Jáchymovské radiové lázně v minulosti a přítomnosti*, Diana Řeháčková, p. 6; Muzeum Královská mincovna Jáchymov, signage viewed 2017; Roger F. Robison (2015) *Mining and Selling Radium and Uranium*, Springer, p. 17.
3. Řeháčková, *Jáchymovské radiové*, p. 6.
4. Robison, *Mining and Selling Radium and Uranium*, p. 18; Muzeum Královská mincovna Jáchymov, signage viewed 2017.
5. Řeháčková, *Jáchymovské radiové*, p. 6.
6. Robison, *Mining and Selling Radium and Uranium*, Springer, p. 203; David I. Harvie (2005) *Deadly Sunshine: The History and Fatal Legacy of Radium*, Tempus, p. 17.
7. Robison, *Mining and Selling Radium and Uranium*, p. 204.
8. Georgius Agricola (1950) *De Re Metallica*: Translated from the First Latin edition of 1556 by Herbert Clark Hoover and Lou Henry Hoover. Dover Publications. Available at: https://www.gutenberg.org/files/38015/38015-h/38015-h.htm (Accessed 1 May 2021). This translation was made by the future American president Herbert Hoover and his wife Lou Henry. Herbert was a mining engineer and Lou was a Latin scholar and geologist, so pretty much a dream team for this project.

9. Roger F. Robison and Richard F. Mould (2006) 'St Joachimsthal: Pitchblende, Uranium and Radon induced Lung Cancer,' *Journal of Oncology* (Vol 56 No 3), p. 277.
10. Harvie, *Deadly Sunshine*, p. 17.
11. Agricola, *De Re Metallica*; Harvie, *Deadly Sunshine*, p. 18.
12. Harvie, *Deadly Sunshine*, p. 19; J.R. Morgan (1984) 'A History of Pitchblende,' *Atom* (No 329 March), p. 63; Robison, *Mining and Selling Radium and Uranium*, p. 205. Interestingly, these masks, which were really loose veils, were made of lace. They needed to be really fine, and so the mine master's wife, Barbara Uttman (known as 'the benefactress of Erzgebirg'), introduced the tradition of fine lace making in the town in 1561.
13. Robison, *Mining and Selling Radium and Uranium*, p. 204.
14. M. Alice Ottoboni (1991) *The Dose Makes the Poison: A Plain Language Guide to Toxicology*, Van Nostrand Reinhold, p. 30; Harvie, *Deadly Sunshine*, p. 19.
15. Aymon De Lestrange (2018) *Coca Wine: Angelo Mariani's Miraculous Elixir and the Birth of Modern Advertising*, Park Street Press, p. 14.
16. Robison, *Mining and Selling Radium and Uranium*, p. 21; Harvie. *Deadly Sunshine*, p. 20.
17. Robison, *Mining and Selling Radium and Uranium*, p. 55.
18. Eduard Farber (1962) *Great Chemists*, John Wiley & Sons, p. 298; Lennard Bickle (1980) *The Deadly Element: the Story of Uranium*, Macmillan, p. 21.
19. Sir James George Frazer, trans. (1959) *Ovid's Fasti*. William Heinemann Ltd, p. 369.
20. Das Glasperlenspiel Archaeochemie (2011) Antique Non-Roman Glass Mosaic in Italy. Available at: http://www.arche-kurzmann.de/ (Accessed 28 January 2021). A further test in 1963 confirmed the presence of uranium.
21. Ibid. While the mosaic disappeared around the Second World War and the tesserae samples have also since been lost, they were retested on several other occasions. The last, in 1963, confirmed the presence of uranium.
22. F. Peter Lole (1995) 'Uranium Glass in 1817: A Pre-Riedel Record,' *Journal of Glass Studies* (Vol 37), p. 139.
23. Harvie, *Deadly Sunshine*, p. 25.
24. Ibid., p. 140; C.S. Gilbert (1817) *An Historical Survey of the County of Cornwall Vol 1*, J. Congdon: Plymouth, p. 269. Available at: https://babel.hathitrust.org/cgi/pt?id=pst.000032375695&view=1up&seq=317&q1=emerald (Accessed 1 March 2021).

25. Barrie Skelcher (1998) 'Uranium Glass'. Available at: http://www. glassassociation.org.uk/sites/default/files/Uranium_Glass_sample_article. pdf (Accessed 26 January 2021).
26. Donna Strahan (2001) 'Uranium in Glass', *Studies in Conservation* (Vol 46), p. 181; Miloš René (2018) 'History of Uranium Mining in Central Europe,' in Nasser S. Awwad (ed) *Uranium – Safety, Resources, Separation and Thermodynamic Calculation*, IntechOpen, p. 4.
27. Skelcher, 'Uranium Glass'.
28. Ibid.
29. Robison, *Mining and Selling Radium and Uranium*, p. 57; Miloš René (2018) 'History of Uranium Mining in Central Europe', p. 4.
30. Robison, *Mining and Selling Radium and Uranium*, p. 57.
31. Ibid., p. 88.
32. Ibid.
33. Řeháčková, *Jáchymovské radiové*, p. 9; Robison, *Mining and Selling Radium and Uranium*, p. 88.
34. Paul Frame & William Kolb (2002) *Living with Radiation: The First Hundred Years*, Syntec, Inc, p. 145; Donna Strahan (2001) 'Uranium in Glass,' p. 182; Jerzy Kunicki-Goldfinger (2018) 'Uran w szkle historycznym,' *Szkło i Ceramika* (Vol 11), p. 19.
35. Frame & Kolb, *Living with Radiation*, p. 147.
36. Ibid.
37. Frame & Kolb, *Living with Radiation*, p. 147.
38. Harvie, *Deadly Sunshine*, p. 23.
39. *Scientific American* (1847) 'Teeth made of stone,' 11 September (Vol 2, No 51), p. 408.
40. John Woodforde (1983) *The Strange Story of False Teeth*, Routledge & Kegan Paul, p. 28.
41. Ibid., p. 61.
42. Ottoboni, *The Dose Makes the Poison*, p. 31.
43. J.R. Morgan (1984) 'A History of Pitchblende,' p. 64.
44. Harold C. Hodge (1973) 'A History of Uranium Poisoning,' in H.C. Hodge, J.N. Stannard and J.B. Hursh (eds.) *Uranium-plutonium Transplutonic Elements*, Springer, p. 5.
45. Hodge (1973) 'A History of Uranium Poisoning,' p. 5.
46. Ibid.
47. Ibid.
48. *Chemist and Druggist* (1895) 'A New Diabetes Remedy,' 31 August, p. 352.
49. Morgan, 'A History of Pitchblende,' p. 65.

50. Harvie, *Deadly Sunshine*, p. 24.

51. *Chemist and Druggist* (1895) 20 April, p. 551.

52. *Chemist and Druggist* (1896) 25 July, p. 113.

53. Ibid., 16 August, p. 285.

54. Ibid., 1 August, p. 228.

55. *Secret Remedies*, p. 78.

56. Ibid.

57. Michael Pritchard (2010) 'The Development and Growth of British Photographic Manufacturing and Retailing 1839–1914,' PhD Thesis, De Montfort University, p. 76. Available at: https://core.ac.uk/download/pdf/228183759. pdf (Accessed 12 January 2021); Daniel Novak (2008) *Realism, Photography and Nineteenth Century Fiction*, Cambridge University Press, p. 11.

58. David G. Präkel (2010) 'Not Fade Away – Lt. Col. Stuart Wortley, the United Association of Photography and the Wothlytype 1863–1867,' Master's Thesis, De Montfort University, p. 1.

59. *The Focal Encyclopaedia of Photography*, (1960) The Macmillan Co, pp. 1191–92; Frame & Kolb, *Living With Radiation*, p. 145.

60. Präkel, 'Not Fade Away,' p. 14.

61. Ibid., p. 11.

62. Ibid., p. 6.

63. Ibid., p. 5.

64. Harvie, *Deadly Sunshine*, p. 23.

65. Präkel, 'Not Fade Away,' p. 52.

66. Ibid., p. 39.

67. Ibid., p. 56.

68. Iwan Rhys Morus (2016) 'Making the Most Beautiful Experiment: Reconstructing Gassiot's Cascade,' in Martin Wills (ed) *Staging Science: Scientific Performance on Street, Stage and Screen*, Palgrave, p. 26.

69. J.L. Lewis and E.J. Wenham (1970) *Radioactivity*, Longman, p. 19.

70. Jeffrey Womack (2020) *Radiation Evangelists: Technology, Therapy and Uncertainty at the Turn of the Century*, University of Pittsburgh Press, p. 9.

71. Robison, *Mining and Selling Radium and Uranium*, p. 32.

72. Womack, *Radiation Evangelists*, p. 9.

73. Robison, *Mining and Selling Radium and Uranium*, p. 32.

74. Deborah Blum, *Ghost Hunters*, Penguin Press, 2007, p. 52.

75. Timothy J. Jorgensen (2016) *Strange Glow: The Story of Radiation*, Princeton University Press, p. 222.

76. Robison, *Mining and Selling Radium and Uranium*, p. 33.

77. Ibid., p. 35.

78. G.I. Brown (2002) *Invisible Rays: A History of Radioactivity*, Sutton Publishing Ltd, p. 9; Farber, *Great Chemists*, p. 278.
79. Peter Broks (1996) *Media Science before the Great War*, Macmillan Press, p. 102.
80. Brown, Invisible Rays, p. 85.
81. Ruth Brecher and Edward M. Brecher (1969) *The Rays: History of Radiology in the United States and Canada*, Lippincott Williams & Wilkins, p. 61; Richard F. Mould (1993) *A Century of X-Rays and Radioactivity in Medicine: With Emphasis on Photographic Records of the Early Years*, CRC Press, p. 39.
82. Claudio Tuniz (2012) *Radioactivity: A Very Short Introduction*, Oxford University Press, p. 61.
83. Joel D. Howell (1995) *Technology in the Hospital: Transforming Patient Care in the Early Twentieth Century*, Johns Hopkins University Press, p. 11.
84. Maria Rentetzi (2011) 'Packaging Radium, Selling Science: Boxes, Bottles and Other Mundane Things in the World of Science,' *Annals of Science* (Vol 68:3), p. 4.
85. *Physics Today* (1996) 'The Discovery of Radium,' February, p. 23; Robison, *Mining and Selling Radium and Uranium*, p. 54.
86. Brown, *Invisible Rays*, p. 12.
87. Eva Hemmungs Wirtén (2015) *Marie Curie: Intellectual Property and Celebrity Culture in an Age of Information*, University of Chicago Press, p. 14.
88. *Birmingham Evening Despatch* (1902) 'X-Ray Slot Machines,' 29 August, p. 7.
89. *Star Tribune* (1896) 3 December, p. 11.
90. *The Electrical Review* (1896) April 17, Vol 38, No 960, p. 511.
91. This is a great film to watch for all Röntgenmaniasts but also notable as it contains one of the first British examples of special effects created by jump cuts.
92. *Guardian* (1913) 1 October, p. 7.
93. Al Look, *U-boom: Uranium on the Colorado Plateau*, Bell Press, 1956, p. 134.

Chapter Two

1. E.N. Jenkins (1979) *Radioactivity: A Science in its Historical and Social Context*, Butterworth & Co, p. 4.
2. Diana Preston (2005) *Before the Fall-Out: The Human Chain Reaction from Marie Curie to Hiroshima*, Doubleday, 2005, p. 22.
3. Ibid., p. 24.
4. Tuniz, *Radioactivity*, p. 7.
5. As found in nature, uranium contains three isotopes: U^{238} (99.27 per cent), U^{235} (0.72 per cent) and U^{234} (0.006 per cent).

6. M. Rayner-Canham and G. Rayner-Canham (1997) 'And Some Other Women of the British Group,' in Marelene F. Rayner-Canham and Geoffrey W. Rayner-Canham (eds.) *A Devotion to their Science: Pioneer Women of Radioactivity*, McGill-Queen's University Press, p. 160; Frederick Soddy and Ruth Pirret (1910) 'The Ratio Between Uranium and Radium in Minerals' *Philosophical Magazine*, (Series 6, Vol 20), p. 349.

7. J. Hudson (2019) 'Dr Margaret Todd and the Introduction of the term "Isotope",' in A. Lykknes and B Van Tiggelen (eds.) *Women in their Element: Selected Women's Contributions to the Periodic System*, World Scientific Publishing, p. 281; Brown, *Invisible Rays: A History of Radioactivity*, p. 63.

8. Brown, *Invisible Rays*, p. 64.

9. Hudson (2019) 'Dr Margaret Todd and the Introduction of the term "Isotope",' p. 280.

10. L. Badash (1978) 'Radium, Radioactivity, and the Popularity of Scientific Discovery,' Proceedings of the American Philosophical Society (Vol 122 No 3), p. 147.

11. Robison, *Mining and Selling Radium and Uranium*, p. 88.

12. Craig Nelson (2014) *The Age of Radiance: The Epic Rise and Dramatic Fall of the Atomic Era*, Scribner Book Company, p. 36.

13. Brynne Voyles, *Wastelanding*, p. vii.

14. Ibid., p. ix.

15. Tom Zoellner (2009) *Uranium: War, Energy, and the Rock that Shaped the World*, Viking, p. 52; Eichstaedt, *If You Poison Us*, p. 9.

16. Eichstaedt, *If You Poison Us*, pp. 9–12.

17. Ibid., p. 9; Brynne Voyles, *Wastelanding*, p. x.

18. Malley, *Radioactivity*, p. 164; Womack, *Radiation Evangelists*, p. 163; Robison, *Mining and Selling Radium and Uranium*, p. 121; Harvie, *Deadly Sunshine*, p. 52.

19. Womack, *Radiation Evangelists*, p. 163; Judy Pasternak (2010) *Yellow Dirt: An American Story of a Poisoned Land and a People Betrayed*, Free Press, p. 28.

20. Pasternak, *Yellow Dirt*, p. 28; Eichstaedt, *If You Poison Us*, p. 12; Robison, *Mining and Selling Radium and Uranium*, p. 130.

21. Robison, *Mining and Selling Radium and Uranium*, p. 130.

22. Ibid.

23. *The Horseless Age* (1907) 'Ford Talks,' 20 March, p. 57.

24. Brynne Voyles, *Wastelanding*, p. 1; Harvie, *Deadly Sunshine*, p. 67.

25. Robison, *Mining and Selling Radium and Uranium*, p. 122.

26. M. Rentetzi (2008) 'The U.S Radium Industry: Industrial In-house Research and the Commercialization of Science,' Minerva (Vol 46), p. 440.

27. Robison, *Mining and Selling Radium and Uranium*, p. 124.

28. Eichstaedt, *If You Poison Us*, p. 11.

29. Malley, *Radioactivity*, p. 164.
30. Eichstaedt, *If You Poison Us*, p. 11.
31. Robison, *Mining and Selling Radium and Uranium*, p. 133.
32. Ibid.
33. Jorgensen, *Strange Glow: The Story of Radiation*, p. 127; Harvie, *Deadly Sunshine*, p. 84.
34. Robison, *Mining and Selling Radium and Uranium*, p. 132; Jorgensen, *Strange Glow*, p. 127; Eichstaedt, *If You Poison Us*, p. 12.
35. Jorgensen, *Strange Glow*, p. 127; Eichstaedt, *If You Poison Us*, p. 12; Robison, *Mining and Selling Radium and Uranium*, p. 133.
36. Robison, *Mining and Selling Radium and Uranium*, pp. 133–34.
37. Eichstaedt, *If You Poison Us*, p. 12.
38. Robison, *Mining and Selling Radium and Uranium*, p. 135.
39. Ibid.; Jorgensen, *Strange Glow*, p. 127.
40. Robison, *Mining and Selling Radium and Uranium*, p. 138.
41. Jorgensen, *Strange Glow*, p. 130; Robison, *Mining and Selling Radium and Uranium*, p. 139.
42. Lucy Jane Santos (2020) *Half Lives: The Unlikely History of Radium*, Icon Books, p. 155.
43. D. Hogarth (2014) 'Robert Richard Sharp 1881–1960: Discoverer of the Shinkolobwe Radium-Uranium Ore Bodies,' *The Journal of the Society for the History of Discoveries* (Vol 46, Issue 1), p. 34.
44. Robert Sharp (1956) *Early Days in Katanga*, Rhodesian Printers, p. 4.
45. D. Hogarth. (2014) 'Robert Richard Sharp 1881–1960' p.37; Zoellner, *Uranium*, p. 5.
46. Harvie, *Deadly Sunshine*, p. 142.
47. Sharp, *Early Days in Katanga*, p. 142.
48. Hogarth (2014) 'Robert Richard Sharp 1881–1960,' p.38.
49. Harvie, *Deadly Sunshine*, p. 142.
50. Zoellner, *Uranium*, p. 43; Badash, *Radioactivity in America*, p. 150.
51. Robison, *Mining and Selling Radium and Uranium*, p. 183.
52. Ibid.
53. Zoellner, *Uranium*, p. 43; Eichstaedt, *If You Poison Us*, p. 20.
54. Ibid., p. 183.
55. Ibid.; Badash, *Radioactivity in America*, p. 150.
56. Ibid.

Chapter Three

1. Malley, *Radioactivity*, p. 113.
2. David Wilson (2004) *Rutherford: Simple Genius*, MIT Press, p. 291.

3. Malley, *Radioactivity*, p. 115.
4. E. Rutherford (1911) 'The Scattering of α and β Particles by Matter and the Structure of the Atom,' *Philosophical Magazine* (Series 6, Vol 21) May, pp. 669–88.
5. Malley, *Radioactivity*, p. 119.
6. *Time* (1932) 'Science: Neutron,' 7 March. Available at: https://content.time.com/time/subscriber/article/0,33009,743285-1,00.html (Accessed 8 March 2023); L.P. Jacquemond (2019) 'Irène Joliot-Curie and the Discovery of "Artificial Radioactivity",' in A. Lykknes and B. Van Tiggelen (eds.) *Women in their Element: Selected Women's Contributions to the Periodic System*, World Scientific Publishing, p. 364.
7. *Time* (1932) 'Science: Neutron,' 7 March.
8. J. Chadwick (1932) 'Possible Existence of a Neutron,' *Nature*, 27 February, p. 312.
9. Brown, *Invisible Rays*, p. 59.
10. *Time* (1932) 'Science: Neutron,' 7 March.
11. APS News (2007) 'This Month in Physics History,' (Vol 16, Number 5). Available at: https://www.aps.org/publications/apsnews/200705/physicshistory.cfm (Accessed 8 March 2023).
12. Amir D. Aczel (2009) *Uranium Wars: The Scientific Rivalry that Created the Nuclear Age*, Palgrave Macmillan, p. 89.
13. Peter Russo (2017) *The Geiger Counter*. Available at: http://large.stanford.edu/courses/2017/ph241/russo2/ (Accessed 25 January 2023).
14. Nelson, *The Age of Radiance*, p. 53.
15. Brown, *Invisible Rays*, p. 36.
16. Aczel, *Uranium Wars*, p. 90.
17. E. Joliot and I. Curie (1934) 'Artificial Production of a New Kind of Radio-Element,' *Nature* (133), pp. 201–02.
18. Irène Joliot – Curie Nobel Lecture. 12 December 1935. Available at: https://www.nobelprize.org/prizes/chemistry/1935/joliot-curie/lecture/ (Accessed 9 April 2023).
19. Preston, *Before the Fall-Out*, p. 89.
20. Ibid., p. 81.
21. AIP 'The First Cyclotrons.' Available at: https://history.aip.org/exhibits/lawrence/first.htm (Accessed 10 April 2023); APS News (2003) 'This Month in Physics History: Lawrence and the First Cyclotron' (Vol 12, No 6). Available at: https://www.aps.org/publications/apsnews/200306/history.cfm (Accessed 10 April 2023).
22. AIP 'The First Cyclotrons.'; Preston, *Before the Fall-Out*, p. 82.

23. AIP 'The First Cyclotrons.'
24. Preston, *Before the Fall-Out*, p.116.
25. Lewis Sime, *Lise Meitner*, p. 163; Nelson, *The Age of Radiance*, p. 162.
26. Aczel, *Uranium Wars*, p. 95.
27. F. Habashi (1997) 'Ida Noddack: Proposer of Nuclear Fission,' in Marelene F. Rayner-Canham and Geoffrey W. Rayner-Canham (eds.) *A Devotion to their Science: Pioneer Women of Radioactivity*, McGill-Queen's University Press, p. 222; Leona Marshall Libby (1979) *The Uranium People*, Russak, p. 42.
28. Aczel, *Uranium Wars*, p. 72.
29. Habashi (1997) 'Ida Noddack,' p. 222.
30. Brown, *Invisible Rays*, p. 98; Aczel, *Uranium Wars*, p. 96.
31. Habashi (1997) 'Ida Noddack,' p. 223.
32. S Watkins (1997) 'Lise Meitner: The Foiled Nobelist,' in Marelene F. Rayner-Canham and Geoffrey W. Rayner-Canham (eds.) *A Devotion to their Science: Pioneer Women of Radioactivity*, McGill-Queen's University Press, p. 182.
33. Habashi (1997) 'Ida Noddack,' p. 222.
34. Ibid.; Lewis Sime, *Lise Meitner*, p. 168.
35. Marshall Libby, *The Uranium People*, p. 43.
36. Lewis Sime, *Lise Meitner*, p. 163.
37. Ibid., p. 184.
38. Andrew Norman (2021) *The Amazing Story of Lise Meitner: Escaping the Nazis and Becoming the World's Greatest Physicist*, Pen and Sword, p. 59.
39. Brown, *Invisible Rays*, p. 98.
40. L. Meitner and O. Frisch (1939) 'Disintegration of Uranium by Neutrons: A New Type of Nuclear Reaction,' *Nature* (Vol 143), p. 239.
41. Watkins (1997) 'Lise Meitner,' p. 187.
42. Habashi (1997) 'Ida Noddack,' p. 223; Catherine Caufield (1989) *Multiple Exposures: Chronicles of the Radium Age*, University of Chicago Press, p. 44.
43. William Lanouette with Bela Silard (1994) *Genius in the Shadows: A Biography of Leo Szilard, the Man behind the Bomb*, University of Chicago Press, p. 55.
44. These had first been proposed by German chemist Max Bodenstein.
45. *The World Set Free* was not the first to mention a new type of bomb. Robert Cromie in *The Crack of Doom* from 1895 writes about a scientist who invents a bomb with 'vast stores of etheric energy locked up in the huge atomic warehouse of this planet'. In late 1914, the *Saturday Evening Post* started serialising *The Man Who Rocked the Earth* by Arthur Train and Robert Wood, which forecast horrifying deaths by radiation poisoning.
46. Spencer R. Weart and Gertrud Weiss Szilard (eds.) (1978) *Leo Szilard, His Version of the Facts: Selected Recollections and Correspondence*, MIT Press, p. 19.

47. Marshall Libby, *The Uranium People*, p. 8; Jeff A. Hughes (2003) *The Manhattan Project: Big Science and the Atom Bomb*, Icon Books, p. 46; Aczel, *Uranium Wars*, p. 124.
48. Spencer Weart (1982) 'The Road to Los Alamos,' *Journal de Physique Colloques*, p. 308; Maxwell Leigh Eidinoff (1947) *Atomics for the Millions*, McGraw-Hill, p. 196; John Campbell (1999) *Rutherford: Scientist Supreme*, AAS Publications, p. 490; Preston, *Before the Fall-Out*, p.141.
49. Robison, *Mining and Selling Radium and Uranium*, p. 189.
50. Aczel, *Uranium Wars*, p. 133.
51. Brown, *Invisible Rays*, p. 106.
52. Ibid.; Jonathan Hogg (2016) *British Nuclear Culture: Official and Unofficial Narratives in the Long 20th Century*, Bloomsbury Academic, p. 38.
53. Aczel, *Uranium Wars*, p. 129; Sean F. Johnston (2009) *History of Science: A Beginner's Guide*, Oneworld Publications, p. 135; Peter Harclerode (2000) *Warfare*, Channel 4 Books, p. 29.
54. Eidinoff, *Atomics for the Millions*, p. 165.
55. Lindsey A. Freeman (2015) *Longing for the Bomb: Oak Ridge and Atomic Nostalgia*, The University of North Caroline Press, p. 18.
56. Brown, *Invisible Rays*, p. 108.
57. Ibid., p. 109.
58. Aczel, *Uranium Wars*, p. 127; Robison, *Mining and Selling Radium and Uranium*, p. 189; Kit Chapman (2021) *Superheavy: Making and Breaking the Periodic Table*, Bloomsbury Sigma, p. 44; Brown, *Invisible Rays*, p. 111; Jon Agar (2012) *Science in the 20th Century and Beyond*, Polity Press, p. 285.
59. Robison, *Mining and Selling Radium and Uranium*, p. 189.
60. Ibid.
61. Ibid.
62. Brown, *Invisible Rays*, p. 111.
63. Nelson, *The Age of Radiance*, p. 118.
64. Chapman, *Superheavy*, p.44; Brown, *Invisible Rays*, p. 111; Robison, *Mining and Selling Radium and Uranium*, p. 189; Nelson, *The Age of Radiance*, p. 118.
65. Brown, *Invisible Rays*, p. 111.
66. A. Costandina Titus (2001) *Bombs in the Backyard: Atomic Testing and American Politics*, University of Nevada Press, p. 7.
67. Titus, *Bombs in the Backyard*, p. 7.
68. Ibid.
69. Aczel, *Uranium Wars*, p. 157.
70. Ibid., p. 158.
71. Denise Kiernan (2013) *The Girls of Atomic City: The Untold Story of the Women Who Helped Win World War II*, Atria Books, p. 76.

72. Aczel, *Uranium Wars*, p. 160.

73. Preston, *Before the Fall-Out*, p. 227.

74. Freeman, *Longing for the Bomb*, p. 63; Brown, *Invisible Rays*, p. 114.

75. Eidinoff, *Atomics for the Millions*, p. 162.

76. Ibid., p. 166.

77. Thorium can be used to breed U-233, another fissile isotope of uranium. It was discovered in 1940 at the University of California at Berkeley.

78. Robison, *Mining and Selling Radium and Uranium*, p. 191; Zoellner, *Uranium*, p. 45.

79. Molly Berkemeier, Wyn Q. Bowen, Christopher Hobbs and Matthew Moran (2014) *Governing Uranium in the United Kingdom*. DIIS Report: 02. Available at: https://pure.diis.dk/ws/files/58173/RP2014_02_Uranium_UK_cve_mfl_web.pdf (Accessed 25 April 2023); Jonathan E. Helmreich (1990) 'Belgium, Britain, The United States and Uranium 1952–1959,' *Studia Diplomatica* (Vol 43 No 3), p. 27.

80. Molly Berkemeier, Wyn Q. Bowen, Christopher Hobbs and Matthew Moran. *Governing Uranium in the United Kingdom*. DIIS Report, 2014: 02.

81. Peter Bacon Hales (1997) *Atomic Spaces: Living on the Manhattan Project*, University of Illinois Press, p. 117.

82. Ibid.

83. Ibid.

84. James A. Mahaffey (2009) *Atomic Awakening: A New Look at the History and Future of Nuclear Power*, Pegasus Books, p. 146.

85. Mahaffey, *Atomic Awakening*, p. 146.

86. Ibid.

87. Ibid., p. 147.

88. Bacon Hales, *Atomic Spaces*, p. 134.

89. Eidinoff, *Atomics for the Millions*, p. 170.

90. Freeman, *Longing for the Bomb*, p. 63.

91. Eidinoff, *Atomics for the Millions*, p. 169.

92. Freeman, *Longing for the Bomb*, p. 18.

93. Eidinoff, *Atomics for the Millions*, p. 169; Robert Johnson (2012) *Romancing the Atom: Nuclear Infatuation from the Radium Girls to Fukushima*, Praeger, p. 89.

94. Brown, *Invisible Rays*, p. 119.

95. Eidinoff, *Atomics for the Millions*, p. 174.

96. Brown, *Invisible Rays*, p. 119.

97. Mahaffey, *Atomic Awakening*, p. 169.

98. Aczel, *Uranium Wars*, p. 173.

99. Caufield, *Multiple Exposures*, p. 57.

100. Aczel, *Uranium Wars*, p. 173; Johnson, *Romancing the Atom*, p. 57.

101. This figure is taken from this fascinating article and guidelines for writing about the tragedy. Alex Wellerstein (2020) 'Counting the dead at Hiroshima and Nagasaki,' *Bulletin of the Atomic Scientists*. Available at: https://thebulletin.org/2020/08/counting-the-dead-at-hiroshima-and-nagasaki/ (Accessed 9 May 2023).

102. Atomic Heritage Foundation 'Szilard Petition,' Available at: https://ahf.nuclearmuseum.org/ahf/key-documents/szilard-petition/ (Accessed 1 May 2023).

103. Ruth Brandon (1987) *The Burning Question: The Anti-nuclear Movement Since 1945*, William Heinemann Ltd, p. 7.

104. *Globe Gazette* (1945) 'Atomic,' 27 August, p. 12.

105. *Weekly Call* (1945) 'Atomic Cocktail,' 18 August, p. 1.

106. Nelson, *The Age of Radiance*, p. 220.

107. *Life* magazine (1945) 'Anatomic Bomb,' 3 September, pp.53–54.

108. Transcript available at: https://wamu.org/story/22/08/14/fourteen-august-a-message-for-the-day-of-victory-by-norman-corwin/ (Accessed 20 April 2023). In a neat parallel to Welles' Invasion from Mars radio play that had caused hysteria in 1939, in February 1946 Parisians listened in horror as Radio Paris broadcast a warning 'Atoms from radium used for research in the United States have broken loose. You will see flashes in the sky. You will have spasms all over the body. You will lose your sense of balance and have difficulty in breathing. Lights will go out in the streets and all cars will stop. The atom fury is nearing Paris and the Atlantic Ocean is in turmoil.' It was, of course, just a play and the radio station had to put out announcements later to reassure their listeners. *Daily Mail* (1946) 'Atom Bomb Play Hits Paris,' 5 February, p. 1.

Chapter Four

1. Preston, *Before the Fall-Out*, p. 5.
2. Santos, *Half Lives*, p. 78.
3. Brown, *Invisible Rays*, p. 96.
4. *Popular Science* (1920) 'Dare We Use This Power?,' May, p. 27.
5. *Collier's Weekly* (1940) 'Fast New World: How atomic power will change your life,' 6 July, p. 54.
6. *Popular Mechanics* (1941) 'The Miracle of U-235,' January (Vol 75), p. 262.
7. *The New York Review of Science Fiction* (2017) 'Beyond "Deadline": Three Additional WWII Atomic War Stories and the Office of Censorship.' Available at: https://www.nyrsf.com/2017/06/steve-carper-beyond-

deadline-three-additional-wwii-atomic-war-stories-and-the-office-of-censorship.html (Accessed 18 April 2023).

8. *The New York Review of Science Fiction* (2017) 'Beyond "Deadline": Three Additional WWII Atomic War Stories and the Office of Censorship.'

9. *Astounding Science Fiction* (1944) 'Deadline,' March (Vol 33, No 1) pp. 154–78.

10. *Asimov's Science Fiction* (2003); Reflections: The Cleve Cartmill Affair Part Two.' Available at: https://web.archive.org/web/20141006183638/http://www.asimovs.com/_issue_0311/ref2.shtml (Accessed 19 April 2023).

11. *Astounding Science Fiction* (1940) 'Blowups Happen,' September (Vol 26, No 1) pp. 51–85. At the Heinlein archive at UCSC there is also a limerick supposedly written by Heinlein around 1940. It goes 'An idea in a scientist's cranium ... Evolved the idea that Uranium ... Applied to a tong to the end of his dong ...Would make it bloom like a geranium.' Huge thanks to Samuel Meyer for drawing my attention to this gem.

12. Zoellner, *Uranium*, p. 88.

13. Ibid., p. 89.

14. Literary Hub (2020) 'The New Yorker Article Heard Round the World.' Available at: https://lithub.com/the-new-yorker-article-heard-round-the-world/ (Accessed 2 August 2022).

15. Ibid.

16. Ibid.

17. Ibid., p. 26.

18. The original five commissioners were David E. Lilienthal, Robert F. Bacher, William W. Waymack, Lewis L. Strauss, Sumner T. Pike.

19. *Life* (1946) 'Peacetime Uses of Atomic Energy,' 2 December, p. 97.

20. Titus, *Bombs in the Backyard*, p. 26.

21. J. Samuel Walker (1992) *Containing the Atom: Nuclear Regulation in a Changing Environment, 1963–1971*, University of California Press, p. 3.

22. Titus, *Bombs in the Backyard*, p. 36.

23. 'Atomic Goddess: Rita Hayworth and the Legend of the Bikini Bombshell' (2011). Available at: http://conelrad.blogspot.com/2011/07/atomic-goddess-rita-hayworth-and-legend.html (Accessed 25 Feb 2023).

24. Titus (1983) 'A-Bombs in the Backyard: Southern Nevada Adapts to the Nuclear Age, 1951–63,' *Nevada Historical Society Quarterly* 26, p. 236.

25. Peter Bacon Hales (2014) *Outside the Gates of Eden: The Dream of America from Hiroshima to Now*, University of Chicago Press, p. 18.

26. Titus, *Bombs in the Backyard*, p. 37; Bacon Hales, *Outside the Gates of Eden*, p. 18.

27. Titus, *Bombs in the Backyard*, p. 37.

28. Johnson, *Romancing the Atom*, p. 28.

29. Titus, *Bombs in the Backyard*, p. 38.
30. 'Atomic Goddess: Rita Hayworth and the Legend of the Bikini Bombshell' (2011). This blog tells the great story of how, for so many years the story made by the press that there was the image of Rita Hayworth on the plane couldn't be collaborated but thanks to some masterful detective work it was proven in 2013.
31. 'Atomic Goddess: Rita Hayworth and the Legend of the Bikini Bombshell' (2011).
32. Ibid.
33. Johnson, *Romancing the Atom*, p. 28; Titus, *Bombs in the Backyard*, p. 37.
34. Ibid.
35. The World (1946), 'Admiral Blandy Atomic Playboy Say Critics; Fan Mail Is Heavy,' 27 May, p. 4.
36. Paul S. Boyer (1985) *By the Bomb's Early Light: American Thought and Culture at the Dawn of the Atomic Age*, Pantheon, p. 11.
37. Mahaffey, *Atomic Awakening*, p. 198.
38. Tiffany Webber-Hanchett (2010) 'Bikini,' in Valerie Steele (ed) *The Berg Companion to Fashion*, Berg, p. 77.
39. Webber-Hanchett (2010) 'Bikini,' p. 77.
40. Johnson, *Romancing the Atom*, p. 32.
41. *The New Yorker* (2016) 'America at the Atomic Crossroads.' Available at: https://www.newyorker.com/tech/annals-of-technology/america-at-the-atomic-crossroads (Accessed 4 August 2022).
42. Titus, *Bombs in the Backyard*, p. 40.
43. *The New Yorker* (2016) 'America at the Atomic Crossroads.'
44. 'Atomic Goddess: Rita Hayworth and the Legend of the Bikini Bombshell' (2011).
45. Johnson, *Romancing the Atom*, p. 25; Zoellner, *Uranium*, p. 193; Boyer, *By the Bomb's Early Light*, p. 296.
46. Allan M. Winkler (1993) *Life Under a Cloud: American Anxiety about the Atom*, Oxford University Press, p. 139; Boyer, *By the Bomb's Early Light*, p. 296.
47. International Disarmament Institute 'Public Events.' Available at: https://disarmament.blogs.pace.edu/nyc-nuclear-archive/civil-defense-and-propaganda/public-events/ (Accessed 8 August 2022).
48. Boyer, *By the Bomb's Early Light*, p. 296.
49. International Disarmament Institute 'Public Events'.
50. Freeman, *Longing for the Bomb*, p. 121.
51. Ibid., p. 122.
52. Ibid., p. 124.
53. Ibid.

Chapter Five

1. *Belfast Telegraph* (1946) 1 July, p. 4.
2. The National Archives, Kew. AB 16/66 (Uranium supply for UK civil consumption).
3. Ibid.
4. Ibid.
5. Lucy Jane Santos (2020) Du Barry Talcum Powder. Available at: https://museumofradium.co.uk/du-barry-talcum-powder/ (Accessed 25 April 2023).
6. The National Archives, Kew. AB 16/66 (Uranium supply for U.K civil consumption); William H. Brock (2008) *William Crookes (1832–1919) and the Commercialization of Science*, Ashgate, p. 464.
7. The National Archives, Kew. AB 16/66 (Uranium supply for UK civil consumption).
8. Atomic Heritage Foundation (2014) Combined Development Trust. Available at: https://ahf.nuclearmuseum.org/ahf/history/combined-development-trust/ (Accessed 24 April 2023).
9. *The New Yorker* (1953) 'A Reporter At Large: The Coming Thing,' 21 March, p. 9.
10. The National Archives, Kew. AB 16/66 (Uranium supply for UK civil consumption).
11. Ibid.
12. Ibid.
13. *Los Angeles Times* (1994) 'Home Safety: Vintage Orange Dishware Poses Threat From Radon Gas'. Available at: https://www.latimes.com/archives/la-xpm-1994-04-23-hm-49397-story.html (Accessed 12 May 2021).
14. Ibid.
15. Michael A. Amundson (2011) *Yellowcake Towns: Uranium Mining Communities in the American West*, University of Colorado, p. 21.
16. Ibid.
17. Pasternak, *Yellow Dirt*, p. 54.
18. Ackland, *Making a Real Killing*, p. 33.
19. Ibid., p. 35.
20. Ibid., p. 36.
21. Ibid.
22. Brynne Voyles, *Wastelanding*, p. 59.
23. Amundson, *Yellowcake Towns*, p. 35.
24. Pasternak, *Yellow Dirt*, p. 57.
25. Ibid., pp 58–59.
26. Harry Kursh (1955) *How to Prospect for Uranium*, Fawcett Publications, p. 41.

27. Ibid., p. 41.

28. Eichstaedt, *If You Poison Us*, p. 36.

29. Ibid.

30. Ibid.

31. Amundson, *Yellowcake Towns*, p. 22; Look, *U-boom*, p. 138; Ringholz, *Uranium Frenzy*, p. 74.

32. Amundson, *Yellowcake Towns*, p. 22; Look, *U-boom*, p. 138; Ringholz, *Uranium Frenzy*, p. 74.

33. Ringholz, *Uranium Frenzy*, p. 74; Look, *U-boom*, p. 138.

34. Robert D. Nininger (1954) 'Hunting Uranium around the World,' *The National Geographic Magazine*, October, p. 547.

35. Look, *U-boom*, p. 138.

36. Amundson, *Yellowcake Towns*, p. 22; Pasternak, *Yellow Dirt*, p. 56.

37. Amundson, *Yellowcake Towns*, p. 22.

38. Kursh, *How to Prospect for Uranium*, p. 8.

39. *Life* (1955) 'History's Greatest Metal Hunt,' 23 May, p. 25; Brynne Voyles. *Wastelanding*, p. 63.

40. https://museumofradium.co.uk/uranium-hunters/

41. *Life* (1955) 'History's Greatest Metal Hunt,' 23 May, p. 25.

42. Look, *U-boom*, p. 106.

43. Kursh, *How to Prospect for Uranium*, p. 52.

44. Caufield, *Multiple Exposures: Chronicles of the Radiation Age*, p. 75.

45. *The New Yorker* (1952) 'The Fourth R,' 19 January, p. 19.

46. *Popular Mechanics* (1955) August, p. 145.

47. Johnson, *Romancing the Atom*, p. 21; Brynne Voyles, *Wastelanding*, p. 61.

48. *The New Yorker* (1949) 'Search,' 27 August, p. 19.

49. *Popular Science* (1946) 'How to Hunt for Uranium,' February 1946, pp. 121–23; Time (1949) 'Out Where the Click is Louder,' 18 July, p. 53; *Life* (1955) 'History's Greatest Metal Hunt,' 23 May, pp. 25–35.

50. *Wichita Falls Times* (1955) 'Uranium News Notes,' 24 April, p. 14.

51. Look, *U-boom*, p. 112.

52. Zoellner, *Uranium*, p. 158.

53. Brynne Voyles, *Wastelanding*, p. 61.

54. Raymond W. Taylor (1970) *Uranium Fever: Or, No Talk under $1 Million*. Macmillan, p. 32.

55. *The New Yorker* (1949) 'Search,' 27 August, p. 19.

56. Brynne Voyles, *Wastelanding*, p. 88; Eichstaedt, *If You Poison Us*, p. 39; Zoellner, *Uranium*, p. 147.

57. Zoellner, *Uranium*, p. 135.

58. Taylor, *Uranium Fever*, p. x; Amundson, *Yellowcake Towns*, p. 25.
59. Johnson, *Romancing the Atom*, p. 21.
60. Adeline Pope McConnell (1954) 'Prospectors without Beards,' *Woman's Day*, 10 October, p. 169.
61. Ringholz, *Uranium Frenzy*, p. 85.
62. Zoellner, Uranium, p. 146; Pasternak, *Yellow Dirt*, p. 85.
63. Ringholz, *Uranium Frenzy*, p. 71; Zoellner, *Uranium*, p. 145.
64. Amundson, *Yellowcake Towns*, p. 26.
65. Lancaster New Era (1955) 'Prospecting Grandma Eyes $1 Million Uranium Offer,' 4 May, p. 1.
66. Scott C. Zeman & Michael A. Amundson (2004) *Atomic Culture: How We Learned to Stop Worrying and Love the Bomb*, University Press of Colorado, p. 51.
67. Amundson, *Yellowcake Towns*, p. 54.
68. Ibid.
69. Zeman & Amundson, *Atomic Culture*, p. 51; Amundson, *Yellowcake Towns*, p. 63; Zoellner, p. 133.
70. Amundson, *Yellowcake Towns*, p. 79.
71. Ibid., p. 93; Zeman & Amundson, *Atomic Culture*, p. 51.
72. Amundson, *Yellowcake Towns*, p. 84.
73. *Life* (1954) 'Uranium Makes a Wilder West,' 19 July, p. 14.
74. Zoellner, *Uranium*, p. 159.
75. Kursh, *How to Prospect for Uranium*, p. 127.
76. Look. *U-boom*, p. 160.
77. *Time* (1954) 'High Finance: Pennies for Uranium,' 5 April. Available at: https://content.time.com/time/subscriber/article/0,33009,819793,00.html (Accessed 27 January 2023).
78. Taylor, *Uranium Fever*, p. 242; *Life* (1954) 'Uranium Makes a Wilder West,' 19 July, p. 14.
79. Ibid., p. 159; Amundson, *Yellowcake Towns*, p. 28.
80. *Life* (1954) 'Uranium Makes a Wilder West,' 19 July, p. 14.
81. Look, *U-boom*, p. 157.
82. Kursh, *How to Prospect for Uranium*, p. 127; Zoellner. *Uranium*, p. 159.
83. Amundson, *Yellowcake Towns*, p. 28.
84. Look, *U-boom*, p. 160.
85. Kursh, *How to Prospect for Uranium*, p. 127.
86. *Life* (1954) 'Uranium Makes a Wilder West,' 19 July, p. 14.
87. *The Record* (1956) 'N.J. Broker Is Indicted a Third Time for Fraud,' 4 August, p. 2.
88. *Popular Mechanics* (1958) 'Beware of Lunar Uranium Stock,' April, p. 78.
89. Taylor, *Uranium Fever*, p. 167.

90. Gabrielle Hecht (2014) *Being Nuclear: Africans and the Global Uranium Trade.* MIT Press, p. 177.
91. Taylor, *Uranium Fever: or, No talk under $1 million*, p. 84.
92. Tim Smolko and Joanna Smolko 2021 *Atomic Tunes: The Cold War in American and British Popular Music,* Indiana University Press, p. 86.
93. Smolko and Smolko, *Atomic Tunes,* p. 86.
94. *The Miami Herald* (1956) 'Radioactive Dirt Cures Ills, Miami Prospector Claims,' 5 February, p. 33.
95. Ibid.
96. *Popular Mechanics* (1960) 'Uranium "Let Him Down" to Success!,' February, p. 33.
97. *Life* (1952) 'Arthritics Seek Cure in Radioactive Mines,' 7 July, p. 22.
98. Archives of the American Medical Association. 0734-10/0758-08 (Rheumatism cures).
99. Robert Wendell Holmes III (2010) 'Substance of the Sun: The Culture History of Radium Medicines in America', PhD Thesis, University of Texas at Austin, p. 252; Barbara Erickson (2000) 'And the People Went to the Caves to be Healed,' in Susanne Bentley, Brad Lucas and Stephen Tchudi (eds.) *Western Futures: Perspectives on the Humanities at the Millennium,* Nevada Humanities Committee, p. 32.
100. Robert Wendell Holmes III (2010) *Substance of the Sun,* p. 33.
101. Archives of the American Medical Association. 0734-10/0758-08 (Rheumatism cures).
102. Robert Wendell Holmes III (2010) *Substance of the Sun,* p. 252. At some point it changed to Free Enterprise Uranium-Radon Mine.
103. Barbara Erickson (2000) 'And the People Went to the Caves to be Healed,' p. 34; Archives of the American Medical Association. 0734-10/0758-08 (Rheumatism cures).
104. Ibid. p. 36.
105. Archives of the American Medical Association. 0734-10/0758-08 (Rheumatism cures).
106. *Life* (1952) 'Arthritics Seek Cure in Radioactive Mines,' 7 July, p. 24.
107. *Time* (1952) 'Medicine: Mind, Body & Mines,' 7 July, Available at: https://content.time.com/time/subscriber/article/0,33009,888728-1,00.html (Accessed 9 April 2023).
108. Archives of the American Medical Association. 0734-10/0758-08 (Rheumatism cures).
109. Claudia Clark (1997) *Radium Girls: Women and Industrial Health Reform, 1910–1935,* University of North Carolina Press, p. 51.

110. Postcard in author's own collection.

111. *Life* (1955) 'Deep in the Dirt of Texas, 24 October, p. 57.

112. *Abilene Reporter-News* (1955) 'Uranium Trail Draws Thousands Seeking Relief From Aches, Pains,' 4 September, p. 1.

113. Ibid.

114. *Abilene Reporter-News* (1955), 11 September 1955, p. 8.

115. *Southern Illinoisan* (1955) 'Uranium Sitter Set,' 7 October, p. 3.

116. *Abilene Reporter-News* (1955), 11 September 1955, p. 8.

117. *Wichita Falls Times* (1955) 'Visiting with Julie,' 20 November, p. 30.

118. *Pampa Daily News* (1956) 'First Time in Years My Feet Haven't Hurt When I Walked,' 9 January, p. 7.

119. *Albuquerque Tribune* (1955) 'Sand Treatment Offered Here at Uranium House,' 29 October, p. 1.

120. *Miami Daily News* (1955) 'Uranium Clinic Probed,' 30 July, p. 1.

121. Ibid.

122. James Harvey Young (1972) *The Toadstool Millionaires*, Princeton University Press, p. 257.

123. Kursh, *How to Prospect for Uranium*, p. 11.

124. ORAU Museum of Radiation and Radioactivity 'Torbena Jar (ca 1929–1931)'. Available at: https://www.orau.org/health-physics-museum/collection/radioactive-quack-cures/jars/torbena-jar.html (Accessed 23 January 2023).

125. Trademark 225,935. US Patent Office. Filed 29 March 1927.

126. Archives of the American Medical Association. 0748-15 (Jackson Uranium Company, Wonderpad).

127. *Fort Worth Star-Telegram* (1955) 'Uranium "Health" Promoter Loses Product in U.S. Seizure,' 4 October, p. 19.

128. ORAU Museum of Radiation and Radioactivity 'Cosmos Radioactive Pad (ca 1950–1956)'. Available at: https://www.orau.org/health-physics-museum/collection/radioactive-quack-cures/radioactive-pads/cosmos-radioactive-pad.html (Accessed 23 January 2023).

129. Santos, *Half Lives*, p. 204.

130. Archives of the American Medical Association. 0748-15 (Jackson Uranium Company, Wonderpad).

131. Ibid.

132. Ibid.

133. Ibid.

134. Food, Drug and Cosmetic Act Notice of Judgment list. Available at: https://collections.nlm.nih.gov/ext/fdanj/ddnj/cases/ddnj05289/ddnj05289.pdf (Accessed 15 March 2023).

135. *The Miami Herald* (1957) 'Sitting House Operators Take Issue With Holle,' 15 January, p. 29.
136. *The Odessa American* (1955) 'Sitting House Operators Take Issue With Holle,' p. 13.
137. Creager, *Life Atomic*, p. 24.
138. F. Barrows Colton (1954) 'Man's New Servant, the Friendly Atom,' *The National Geographic Magazine*, January, p. 77.
139. Ibid.
140. Creager, *Life Atomic*, p. 84.
141. Brown, *Invisible Rays*, p. 163.
142. Edward R. Landa, 'Buried Treasure to Buried Waste: The Rise and Fall of the Radium Industry,' *Colorado School of Mines Quarterly*, 1987, p. 42; Jorgensen, *Strange Glow: The Story of Radiation*, p. 89.
143. Creager, *Life Atomic*, p. 27.
144. *Popular Mechanics* (1948) 'Radioactive Gold Aids Cancer War,' February, p. 95.
145. Creager, *Life Atomic*, p. 36.
146. *Democrat and Chronicle* (1935) 'Has Practical Possibilities,' 6 September, p. 16.
147. Creager, *Life Atomic*, p. 33.
148. Creager, *Life Atomic*, p. 3.
149. *Life* (1950) 'Atomic Progress,' 1 January, pp. 23–35.
150. F. Barrows Colton (1954) 'Man's New Servant, the Friendly Atom,' *The National Geographic Magazine*, January, p. 83.
151. Laura Fermi. (1957) *Atoms for the World: United States Participation in the Conference on the Peaceful Uses of Atomic Energy*, University of Chicago Press, p. 51.
152. *Life* (1954) Bendix Home Appliances advert, 11 October, p. 173.
153. *Life* (1960) Absorbine Jr advert, 22 August, p. 76.
154. *Cosmetics and Skin* 'Salon Cold Cream cleanses 2 ½ times more thoroughly.' Available at: http://www.cosmeticsandskin.com/booklets/gray-cold.php (Accessed 21 July 2021).
155. Ibid.
156. Ibid.
157. 'Dorothy Gray Salon Cold Cream' Available at: https://www.youtube.com/watch?v=_HylVqmJnIc&t=21s
158. *New York Times* (2010) 'John Shepherd-Barron, Developer of the A.T.M, Dies at 84,' 20 May. Available at: https://www.nytimes.com/2010/05/21/business/global/21barron.html (Accessed 11 May 2023).

Chapter Six

1. Titus, *Bombs in the Backyard*, p. 55; Alice Buck, The Atomic Energy Commission, July 1983, p. 11. Available at https://www.energy.gov/management/articles/history-atomic-energy-commission (Accessed 1 June 2023).
2. Titus, *Bombs in the Backyard*, p. 71.
3. Ibid., p. 55.
4. Barbara Rose Johnston (2007) *Half-lives and Half-truths: Confronting the Radioactive Legacies of the Cold War*, School for Advanced Research Press, p. 194.
5. Larry D. Gragg (2019) *Becoming America's Playground: Las Vegas in the 1950s*, University of Oklahoma Press, p. 147.
6. Titus (1983) 'A-Bombs in the Backyard: Southern Nevada Adapts to the Nuclear Age, 1951–63,' p. 239; Gragg, *Becoming America's Playground*, p. 147.
7. Titus, *Bombs in the Backyard*, p. 60.
8. Madeline Auerbach (2017) 'Dancing into the Mushroom Cloud: The Idealization of the Atomic Bomb in 1950s Las Vegas,' BA Thesis, Georgetown University, p. 37.
9. Edward E. Baldwin (1996) 'Las Vegas in Popular Culture,' PhD Thesis, University of Nevada, Las Vegas, p. 9.
10. Auerbach (2017) *Dancing into the Mushroom Cloud*, p. 43.
11. Fred Pearce (2018) *Fallout: Disasters, Lies and the Legacy of the Nuclear Age*, Beacon Press, p. 24.
12. Daniel Lang (1952) 'Our Far-Flung Correspondent: Blackjack and Flashes,' *The New Yorker*, September 20, p. 104.
13. Johnston, *Half-lives and Half-truths*, p. 195; Gragg, *Becoming America's Playground*, p. 149
14. Daniel Lang (1952) 'Our Far-Flung Correspondent: Blackjack and Flashes,' *The New Yorker*, September 20, p. 105.
15. Auerbach, *'Dancing into the Mushroom Cloud'*, p. 38.
16. Ibid., p. 31.
17. Gragg, *Becoming America's Playground*, p. 157.
18. Ibid.; Titus, *Bombs in the Backyard*, p. 93.
19. Gragg, *Becoming America's Playground*, p. 156.
20. Titus, *Bombs in the Backyard*, p. 93.
21. Gragg, *Becoming America's Playground*, p. 157; Titus, *Bombs in the Backyard*, p. 94.
22. Gragg, *Becoming America's Playground*, p. 157.

23. Titus, *Bombs in the Backyard*, p. 94.
24. Daniel Lang (1952) 'Our Far-Flung Correspondent: Blackjack and Flashes,' *The New Yorker*, 20 September, p. 101.
25. Ibid., p. 107.
26. Smolko and Smolko, *Atomic Tunes*, p. 130.
27. Don English quoted in Auerbach, *'Dancing into the Mushroom Cloud'*, p. 49.
28. Gragg, *Becoming America's Playground*, p. 107.
29. Daniel Lang (1952) 'Our Far-Flung Correspondent: Blackjack and Flashes,' *The New Yorker*, 20 September, p. 100.
30. Titus, *Bombs in the Backyard*, p. 95.
31. Ibid.
32. Titus, 'A-Bombs in the Backyard,' p. 249.
33. Titus, *Bombs in the Backyard*, p. 95.
34. Bacon Hales, *Outside the Gates of Eden*, p. 143; Gragg, *Becoming America's Playground*, p. 148.
35. Titus, *Bombs in the Backyard*, p. 96.
36. Titus, 'A-Bombs in the Backyard,' p. 242; Joanna Bourke (2015) *Fear: A Cultural History*, Virago, p. 284.
37. Smolko and Smolko, *Atomic Tunes*, p. 116.
38. Zoellner, *Uranium*, p. 99.
39. Smolko and Smolko, *Atomic Tunes*, p. 142.
40. *Life* (1961) 'Fallout Shelters: A New Urgency, Big Things to Do and What You Must Learn,' 15 September, p. 94.
41. Smolko and Smolko, *Atomic Tunes*, p. 116.
42. Ibid.
43. Advert in author's own collection.
44. Kenneth D. Rose (2001) *One Nation Underground: The Fallout Shelter in American Culture*, New York University Press, p. 1.
45. Zoellner, *Uranium*, p. 100; Rose, *One Nation Underground*, p. 10.
46. *Los Angeles Times* (1954) 'Uranium Fever,' 25 November, p. 55.
47. Smolko and Smolko, *Atomic Tunes*, p. 114.
48. *Life* (1950) 20 November, p. 138.
49. Richard Rhodes (2018) *Energy: A Human History*, Simon & Schuster, p. 317.
50. Titus, *Bombs in the Backyard*, p. 86.
51. Titus, 'A-Bombs in the Backyard,' p. 245; Zeman & Amundson, *Atomic Culture: How we Learned to Stop Worrying and Love the Bomb*, p. 4.
52. Johnston, *Half-lives and Half-truths*, p. 214; Johnson, *Romancing the Atom*, p. 38.

53. Johnston, *Half-lives and Half-truths*, p. 214; Johnson, *Romancing the Atom*, p. 38.
54. Johnson, *Romancing the Atom*, p. 38.
55. Johnston, *Half-lives and Half-truths*, p. 214.
56. Creager, *Life Atomic*, p. 147.
57. Caufield, *Multiple Exposures*, p. 123.
58. Zoellner, *Uranium*, p. 94; Pearce, *Fallout*, p. 26.
59. Rhodes, *Energy*, p. 317; Jerry M. Cuttler (2015) 'Nuclear Energy and the LNT Hypothesis of Radiation Carcinogenesis,' in Shizuyo Sutou (ed) *Fukushima Nuclear Accident*, Nova Science Publishers, p. 29.
60. Jerry M. Cuttler 'Nuclear Energy and the LNT Hypothesis,' p. 45.
61. Jerry M. Cuttler 'Nuclear Energy and the LNT Hypothesis,' p. 30; Gragg, *Becoming America's Playground*, p. 153.
62. *Playboy* (1959) 'The Contaminators,' October (Vol 6 No 10), p. 38.
63. J.D. Ratcliff (1950) 'Are You Getting Too Much X-Ray?', *Red Book* (Vol 95, Issue 1), May, p. 53.
64. ORAU Museum of Radiation and Radioactivity. Shoe Fitting Fluoroscope (ca 1930–1940). Available at: https://www.orau.org/health-physics-museum/collection/shoe-fitting-fluoroscope/index.html (Accessed 16 May 2023).
65. Teresa Riordan (2004) *Inventing Beauty: A History of the Innovations that Have Made Us Beautiful*, Broadway Books, p. 31.
66. ORAU Museum of Radiation and Radioactivity. Shoe Fitting Fluoroscope (ca 1930–1940).
67. Ibid.
68. Bob McCoy (2000) *Quack!: Tales of Medical Fraud from the Museum of Questionable Medical Devices*, Santa Monica Press, p. 132.
69. Gragg, *Becoming America's Playground*, p. 153.
70. Ibid.
71. Titus, *Bombs in the Backyard*, p. 99.
72. Titus, 'A-Bombs in the Backyard,' p. 245; Allan M. Winkler (1993) *Life under a Cloud: American Anxiety about the Atom*, Oxford University Press, p. 105
73. Winkler, *Life under a Cloud*, p. 105.
74. Ibid., p. 106.
75. Ibid.; Titus, 'A-Bombs in the Backyard,' p. 245.
76. Smolko and Smolko, *Atomic Tunes*, p. 53.
77. Campaign for Nuclear Disarmament 'History of the Symbol,' Available at: https://cnduk.org/the-symbol/ (Accessed 12 December 2022).

Chapter Seven

1. Piles of the Future Review (1944). Available at: https://technicalreports. ornl.gov/1944/3445605715063.pdf; Prospectus on Nucleonics (1944). Available at: https://atomicinsights.com/wp-content/uploads/Prospectus-on-Nucleonics.pdf (Accessed 12 June 2023).
2. Prospectus on Nucleonics (1944).
3. Rhodes, *Energy*, p. 277.
4. Gerald Wendt (1945) *The Atomic Age Opens*, The World Publishing Company, p. 175.
5. John O'Neill (1945) *Almighty Atom: The Real Story of Atomic Energy*, Ives Washburn Inc, p. 64.
6. *Science Illustrated* (1949) 'The Coming Atomic Age Offers Some Awesome Possibilities,' February pp. 12–13.
7. Dr R.M. Langer. 'The Miracle of U-235,' *Popular Mechanics*, January 1941, p. 2.
8. Boyer. *By the Bomb's Early Light*, p. 112.
9. *Democrat and Chronicle* (1945) '"Atomic Car" Inventor Claims Ability to Propel Any Vehicle,' 19 November, p. 7.
10. Ibid.
11. Ibid.
12. *Guardian* (1945) 'No Test of the "Atomic Car",' 30 November, p. 6.
13. *Cairns Post* (1946) '"Atomic Car" Hoax,' 22 July, p. 3.
14. *Evening Chronicle* (1945) 'Last years of petrol driven car,' 7 December, p. 5.
15. Auto Blog (2014) 'Nuclear-powered concept cars from the Atomic Age,' Available at: https://www.autoblog.com/2014/07/17/nuclear-powered-atomic-age-classic-cars/?guccounter=1 (Accessed 6 May 2023).
16. CNN Business (2019) 'Electric cars have been around since before the US Civil War,' Available at: https://edition.cnn.com/interactive/2019/07/business/electric-car-timeline/index.html (Accessed 19 January 2023).
17. Ethical Net 'The Electric Car Problem,' Available at: https://ethical.net/transport/we-had-electric-cars-in-1900-then-this-happened/ (Accessed 17 January 2023).
18. Advert in author's own collection.
19. Boyer, *By the Bomb's Early Light*, p. 115.
20. Walker, *Containing the Atom*, 1963–1971, p. 99.
21. What is Nuclear (2020), 'Nuclear Reactor Development History,' Available at: https://whatisnuclear.com/reactor_history.html (Accessed 5 June 2023).
22. Fermi, *Atoms for the World*, p. 5.
23. Ibid.

24. Winkler, *Life under a Cloud*, p. 145.
25. Fermi, *Atoms for the World*, p. 6.
26. Buck, *The Atomic Energy Commission*, p. 6; Rhodes, *Energy*, p. 284; Winkler, *Life under a Cloud*, p. 145.
27. Fermi, *Atoms for the World*, p. 7; Titus, *Bombs in the Backyard*, p. 79.
28. Titus, *Bombs in the Backyard*, p. 77.
29. San Francisco Maritime National Park Association (2020), 'Tour of the NS Savannah,' Available at: https://maritime.org/tour/savannah/ (Accessed 3 July 2023).
30. Rhodes, *Energy*, p. 284.
31. Fermi, *Atoms for the World*, p. 27.
32. Ibid., p. 286.
33. Buck, *The Atomic Energy Commission*, p. 7.
34. Walker, *Containing the Atom*, p. 9.
35. Buck, *The Atomic Energy Commission*, p. 7.
36. Ibid.; Winkler, *Life under a Cloud*, p. 149; Walker, *Containing the Atom*, p. 9.
37. Walker. *Containing the Atom*, p. 9.
38. Mahaffey, *Atomic Awakening*, p. 206; Rhodes. *Energy*, p. 285.
39. Anthony Burke (2017) *Uranium*, Polity Press, p. 24.
40. Aczel, *UraniumWars*, p. 213; Hunter Davies (2012) *Sellafield Stories*, Constable, p.17.
41. Caufield, *Multiple Exposures*, p. 150.
42. Shippingport Atomic Power Station: A National Historic Mechanical Engineering Landmark, 1980, p.4.
43. Mahaffey, *Atomic Awakening*, p. 229.
44. Shippingport Atomic Power Station, p.5.
45. Buck, *The Atomic Energy Commission*, p. 7.
46. Walker, *Containing the Atom*, p. 19.
47. What is Nuclear (2020), 'Nuclear Reactor Development History.'
48. F. Barrows Colton (1954) 'Man's New Servant, the Friendly Atom,' *The National Geographic Magazine*, January, p. 88.
49. Ibid.
50. Buck, *The Atomic Energy Commission*, p. 6.
51. Walker, *Containing the Atom*, p. 20.
52. Buck, *The Atomic Energy Commission*, p. 7; Mahaffey, *Atomic Awakening*, p. 229.
53. Mahaffey, *Atomic Awakening*, p. 229; Rhodes, *Energy*, p. 290.
54. Rhodes. *Energy*, p. 289; Aczel, *Uranium Wars*, p. 214; Mahaffey, *Atomic Awakening*, p. 229.
55. Shippingport Atomic Power Station, p.10.

56. *Hinton Daily News* (1957) 'Dispelling the Gloom,' 11 December, p. 4; *The Times* (1957) '1st Atomic Electricity Turned Out,' 18 December, p. 3; *The Philadelphia Inquirer* (1957) 'We Enter the Age of the Peaceful Atom,' 20 December, p. 28.
57. Ibid., p. 21.
58. Ibid., p. 21.
59. Ibid., p. 37.
60. Alvin Martin Weinberg (1985) *The Second Nuclear Era: A New Start for Nuclear Power*, Praeger, p. 88.
61. Weinberg, *The Second Nuclear Era*, p. 102; Buck, *The Atomic Energy Commission*, p. 7.
62. Lucy Jane Santos (2024) '"The Triumph of the New Over the Old": Electric Restaurants, Health and Modernity,' in Lauren Alex O'Hagan and Göran Eriksson (eds.) *Food Marketing and Selling Healthy Lifestyles with Science: Transhistorical Perspectives*. Forthcoming.
63. Alison Boyle (2019) '"Banishing the Atom Pile Bogy": Exhibiting Britain's First Nuclear Reactor,' *Centaurus* (Vol 61, Issue 1–2), p. 17.
64. Ibid., p.19.
65. Dennis Pohl (2021) 'Uranium Exposed at Expo 58: The Colonial Agenda behind the Peaceful Atom,' *History and Technology* (Vol 37), p. 172.
66. Ibid., p. 185.
67. Ibid., p. 183.
68. Card in author's own collection.
69. Walker, *Containing the Atom*, p. 12.
70. Rhodes, *Energy*, p. 313; Walker, *Containing the Atom*, p. 12.
71. Walker, *Containing the Atom*, p. 12; Mahaffey, *Atomic Awakening*, p. 303.
72. Jerry Brown and Rinaldo Brutoco (1997) *Profiles in Power: The Antinuclear Movement and the Dawn of the Solar Age*, Twayne Publishers, p. 7.
73. Howard Margolis (1962) 'Atomic Power: Cinderella Is Slipping Back into the Kitchen,' *Science*, (136) p. 244; 136 (20 April 1962), p. 244.
74. Winkler, *Life under a Cloud*, p. 151.
75. Ibid., p. 151.
76. Walker, *Containing the Atom*, p. 16.
77. Winkler, *Life under a Cloud*, p. 151.
78. Buck, *The Atomic Energy Commission*, p. 11; Walker, *Containing the Atom*, p. 16.
79. Auerbach, *Dancing into the Mushroom Cloud*, p. 68.
80. Buck, *The Atomic Energy Commission*, p. 11.

81. Yamika Herald- Republic (2013) 'It Happened Here: JFK breaks ground for nuclear reactor at Hanford.' Available at: https://www.yakimaherald.com/news/local/it-happened-here-jfk-breaks-ground-for-nuclear-reactor-at-hanford/article_1514bad2-7089-5487-a0ab-40af860922d2.html (Accessed 9 June 2023).

82. Atomic Insights (2013) 'JFK's "Best of the Above" speech at Hanford. Available at: https://atomicinsights.com/jfks-best-speech-hanford-wa-september-26-1963/ (Accessed 12 June 2023).

83. History Link (2013) 'President Kennedy participates in groundbreaking ceremonies for construction of N Reactor at Hanford.'

84. *Yamika Herald-Republic* (2013) 'It Happened Here: JFK breaks ground for nuclear reactor at Hanford.'

85. Walker, *Containing the Atom*, p. 28.

86. Rhodes, *Energy*, p. 314.

87. Walker, *Containing the Atom*, p. 30; Winkler, *Life under a Cloud*, p. 151.

88. Buck, *The Atomic Energy Commission*, p. 11; Thomas Raymond Wellock, *Critical Masses: Opposition to Nuclear Power in California 1958–1978*, The University of Wisconsin Press, 1998, p. 72.

89. Hecht, *Being Nuclear: Africans and the Global Uranium Trade*, p. 60.

90. Johnston, *Half-lives and Half-truths*, p. 118; Buck, *The Atomic Energy Commission*, p. 11; Walker, *Containing the Atom*, p. 24.

91. Ibid., p. 228.

92. Taylor, *Uranium Fever*, p. 103.

93. Taylor, *Uranium Fever*, p. 104.

94. Pasternak, *Yellow Dirt*, p. 104.

95. Zoellner, *Uranium*, p. 171; Taylor, *Uranium Fever*, p. 252; Eichstaedt, *If You Poison Us*, p. 37.

96. Johnston, *Half-lives and Half-truths*, p. 3.

97. Eichstaedt, *If You Poison Us*, p. 36.

98. Taylor, *Uranium Fever*, p. 104.

99. Ibid.; Zoellner, *Uranium*, p. 171.

100. Taylor, *Uranium Fever*, p. 252.

101. Johnson, *Romancing the Atom*, p. 59.

102. Agricola, *De Re Metallica*.

103. Johnston, *Half-lives and Half-truths*, p. 119; Eichstaedt, *If You Poison Us*, p. 56.

104. Eichstaedt, *If You Poison Us*, p. 54; Zoellner, *Uranium*, p. 169.

105. Johnston, *Half-lives and Half-truths*, p. 120.

106. Ibid., p. 121; Pasternak, *Yellow Dirt*, p. 92; Brynne Voyles, *Wastelanding*, p. 110.

107. Pasternak, *Yellow Dirt*, p. 94.
108. Walker, *Containing the Atom*, p. 241.
109. Ibid., p. 243.
110. Ibid., p. 247.
111. Ibid., p. 31.
112. Wellock, *Critical Masses*, p. 72; Weinberg, *The Second Nuclear Era*, p. 51.
113. Winkler, *Life under a Cloud*, p. 160.
114. Walker, *Containing the Atom*, p. 1.
115. Wellock, *Critical Masses*, p. 74.
116. Mahaffey, *Atomic Awakening*, p. 303; Walker, *Containing the Atom*, p. 388.
117. Walker, *Containing the Atom*, p. 91.
118. Ibid., p. 394.
119. Wellock, *Critical Masses*, p. 88; Winkler, *Life under a Cloud*, p. 160.
120. Winkler, *Life under a Cloud*, p. 160; Walker, *Containing the Atom*, p. 267.
121. Wellock, *Critical Masses*, p. 88; Winkler, *Life under a Cloud*, p. 160.
122. Walker, *Containing the Atom*, p. 33.
123. Taylor, *Uranium Fever*, p. 313.
124. Walker, *Containing the Atom*, p. 31.
125. Buck, *The Atomic Energy Commission*, p. 16; Winkler, *Life under a Cloud*, p. 160.
126. Buck, *The Atomic Energy Commission*, p. 16.
127. Ibid., p. 16.
128. Wellock, *Critical Masses*, p. 89.
129. Walker, *Containing the Atom*, p. 397.
130. Ibid.
131. Wellock, *Critical Masses*, p. 107.
132. Ibid., p. 110.
133. Walker, *Containing the Atom*, p. 408.
134. Winkler, *Life under a Cloud*, p. 153; Walker. *Containing the Atom*, p. 408.
135. Winkler, *Life under a Cloud*, p. 153.
136. Buck, *The Atomic Energy Commission*, p. 18.
137. Mahaffey, *Atomic Awakening*, p. xvii.
138. Ibid.
139. Walker, *Containing the Atom*, p. 408.
140. Wellock, *Critical Masses*, p. 95.
141. Christian Joppke (1993) *Mobilizing Against Nuclear Energy: A Comparison of Germany and the United States*, University of California Press, p. 31.
142. Marco Giugni (2004) *Social Protest and Policy Change: Ecology, Antinuclear, and Peace Movements in Comparative Perspective*, Rowman & Littlefield, p. 43.

143. Mahaffey, *Atomic Awakening*, p. 311; What is Nuclear 'General History of Nuclear Energy' Available at: https://whatisnuclear.com/history.html (Accessed 5 June 2023).
144. Mahaffey, *Atomic Awakening*, p. 311.
145. Winkler, *Life under a Cloud*, p. 161.
146. Spiegel International (2011) 'The Origin of the Anti-Nuclear Emblem.' Available at: https://www.spiegel.de/international/zeitgeist/the-origin-of-the-anti-nuclear-emblem-we-wanted-a-logo-that-was-cheerful-and-polite-a-773903.html (Accessed 12 December 2022).
147. Weinberg, *The Second Nuclear Era*, p. 18.

Chapter Eight

1. And maybe a little bit because it's just plain rude that you were alive during 'history'.
2. Weinberg, *The Second Nuclear Era*, p. 124. In April 1979, the NRC announced cumulative data from environmental monitoring that included air, water, soil and milk samples from the area around the TMI power plant a week after the accident. The radioactivity in the area thankfully was minimal. The majority of the radiation was contained, as by its design, in the containment structure – the likes of which have been in use since the 1950s in the US.
3. Winkler, *Life under a Cloud*, p. 158.
4. Nuclear Newswire (2022) 'Meltdown: Drama disguised as a documentary,' Available at: https://www.ans.org/news/article-4016/meltdown-drama-disguised-as-a-documentary/ (Accessed 18 October 2022).
5. *Pottsville Republican* (1979) 'TMI: terror unleashed,' 10 April, p. 1; *Pittsburgh Press* (1979) 'Life'll Never Be Same, Nuclear Refugee Says,' 3 April, p. 4; Weinberg, *The Second Nuclear: Era*, p. 126.
6. Johnston, *History of Science*, p. 135; Harclerode, *Warfare*, p. 151; Winkler, *Life under a Cloud*, p. 158.
7. Finis Dunaway (2015) *Seeing Green – The Use and Abuse of American Environmental Images*, University of Chicago Press, p. 124.
8. Weinberg, *The Second Nuclear Era*, p. 126; Winkler, *Life under a Cloud*, p. 158.
9. Weinberg, *The Second Nuclear Era*, p. 131.
10. Ibid., p. 19.
11. Ibid., p. 138.
12. Ibid.
13. Object in author's own collection.
14. Dunaway, *Seeing Green*, p. 141.

15. Mahaffey, *Atomic Awakening*, p. 311.
16. Weinberg, *The Second Nuclear Era*, p. 126.
17. Smolko and Smolko, *Atomic Tunes*, p. 9.
18. Nelson, *The Age of Radiance*, p. 310.
19. Object in author's own collection.
20. Mahaffey, *Atomic Awakening*, p. 310.
21. Ibid.
22. Gwyneth Cravens (2007) *Power to Save the World: The Truth About Nuclear Energy*, Alfred A Knopf, p. 9. It is estimated that the entire spent fuel waste from reactors would fit in fewer than 9,000 dry casks.
23. Mahaffey, *Atomic Awakening*, p. 310.
24. Alice Gibbs (2023) 'Pregnant Woman Poses With "Nuclear Waste" To Prove Point About Radiation,' *Newsweek*, 28 June, Available at: https://www.newsweek.com/pregnant-woman-poses-nuclear-waste-prove-point-about-radiation-idaho-1809500?utm_medium=Social&utm_source=Twitter#Echobox=1687945076 (Accessed 28 June 2023).
25. Ibid.
26. Johnson, *Romancing the Atom*, p. 66.
27. Weinberg, *The Second Nuclear Era*, p. 19.
28. Sigvard Eklund (1981) 'Nuclear Power Development – The Challenge of the 1980s,' IAEA Bulletin (Vol 23 No 3), p. 8.
29. Brown and Brutoco, *Profiles in Power*, p. 31.
30. Aczel. *Uranium Wars*, p. 217; Winkler, *Life under a Cloud*, p. 159.
31. Winkler, *Life under a Cloud*, p. 159.
32. Aczel, *Uranium Wars*, p. 217.
33. Ibid.
34. Mahaffey, *Atomic Awakening*, p. 318.
35. Aczel, *Uranium Wars*, p. 217; Winkler. *Life under a Cloud*, p. 159.
36. Michael Burgan (2018) *Chernobyl Explosion: How a Deadly Nuclear Accident Frightened the World*, Capstone Press, p. 8.
37. Johnston, *Half-lives and Half-truths*, p. 109.
38. Zoellner, *Uranium*, p. 172.
39. Burke, *Uranium*, p. 76; Ariel Gould (2002) 'Sustainable and Ethical Uranium Mining: Opportunities and Challenges,' Good Energy Collective, August 31. Available at: https://www.goodenergycollective.org/policy/sustainable-and-ethical-uranium-mining-opportunities-and-challenges (Accessed 10 May 2023).
40. Aczel, *Uranium Wars*, p. 218; Johnston, *History of Science*, p. 151.

41. Maxwell Irvine 2011 *Nuclear Power: A Very Short Introduction*, Oxford University Press, p. 70.
42. IEA (2019) 'Nuclear Power in a Clean Energy System.' Available at: https://www.iea.org/reports/nuclear-power-in-a-clean-energy-system (Accessed 8 June 2023).
43. Joshua S. Goldstein and Staffan A. Qvist (2020) *A Bright Future: How Some Countries Have Solved Climate Change and the Rest Can Follow*, Hachette Book Group, p. 95.
44. The other reason is racism.
45. Jerry M. Cuttler (2014) 'Remedy for Radiation Fear – Discard the Politicized Science,' Dose Response, 13 March. Available at: https://journals.sagepub.com/doi/full/10.2203/dose-response.13-055.Cuttler (Accessed 5 October 2022).
46. Atomik is sold by the Chernobyl Spirit Company, a social enterprise. At least 75 per cent of profits go to helping rebuild communities in Ukraine first impacted by Chernobyl and then by the war. It is blooming tasty!
47. World Nuclear (2023) 'Plans for New Reactors Worldwide,' Available at: https://world-nuclear.org/information-library/current-and-future-generation/plans-for-new-reactors-worldwide.aspx (Accessed 28 June 2023).
48. World Nuclear (2023) 'Nuclear Power in the World Today.'

INDEX